A Problem-Based Guide to Basic Genetics

Fifth Edition

Donald Cronkite
Hope College

THOMSON

BROOKS/COLE

Australia • Brazil • Canada • Mexico • Singapore • Spain • United Kingdom • United States

Printed in the United States of America

1 2 3 4 5 6 7   11 10 09 08 07

Printer: Thomson/West
Cover Image: © Masterfile Corporation

ISBN-13: 978-0-495-38468-7
ISBN-10: 0-495-38468-2

Thomson Higher Education
10 Davis Drive
Belmont, CA 94002-3098
USA

For more information about our products,
contact us at:
Thomson Learning Academic Resource Center
1-800-423-0563

For permission to use material from this text or
product, submit a request online at
http://www.thomsonrights.com.
Any additional questions about permissions can be
submitted by email to **thomsonrights@thomson.com.**

PREFACE

A PROBLEM-BASED GUIDE TO BASIC GENETICS

This is the fifth edition of the *Problem-Based Guide*. It started its life as a 35-page supplement to the 4th Edition of Brooks/ Cole's *Biology* by Solomon, Berg and Martin in 1996. The first sentence of that first edition still expresses our belief: "Solving genetics problems is not very difficult if you are systematic in your approach." Being systematic and learning to watch for patterns was what the guide offered in 1996 and what it offers today. There are more problems and more aspects of genetics than in that first edition. Past revisions added chapters on chromosomes and linkage and new problems based on recent research. This time three further chapters offer guidance on probability and statistics, population genetics, and quantitative genetics.

We hope this will expand the usefulness to a wider range of readers. Instructors who use the guide can skip these chapters if they are beyond the scope of the students. Nothing in the other chapters depends on these new chapters, but they are there if anyone needs help on these important areas. The new chapters move students into a slightly higher level of math than the others do.

We have praised Mendel for his insight in discerning the mechanism of inheritance in his peas. The new chapters are in this Mendelian tradition. They use simple math and a little reasoning to solve them. After more than a century, thinking in terms of simple behavior patterns of these abstracts factors called genes brings understanding of heredity, not just in peas, but in people too. Mendelian genetics has contributed in important ways to our technology, but not just that. As a form of reasoning Mendelian genetics is an important part of culture.

And so, once again, I thank Johann Gregor Mendel for inspiring the guide and students who have used the guide and told me what they think. Thanks also to two essential people from Brooks/Cole, Peter Adams, the publisher, and Lauren Oliveira, the editor. This edition is as much theirs as mine, each for his or her own reason. Thanks also to Jane Cronkite for being who she is on all my projects.

Donald Cronkite

TABLE OF CONTENTS

A PROBLEM BASED GUIDE TO GENETICS

Preface

Table of Contents

1. Six Systematic Steps for Solving Genetics Problems………………….……..1

2. Going beyond the Six Steps……………………………………..….21

3. Probability and Statistics……………………………………………...29

4. More Complex Situations………………………………………….47

5. Sex Linkage……………………………………………………….61

6. Pedigrees……………………………………………………….69

7. Linkage……………………………………………………….81

8. Linkage Maps………………………………………………….....91

9. Chromosomes…………………………………………………..…100

10. Population Genetics…………………………………………………......119

11. Quantitative Genetics………………………………………………......129

12. A collection of Review Problems……………………………………145

Bibliography……………………………………………………149

Answers to Practice Problems……………………………………….....152

Glossary………………………………………………………201

Chi-Square Table of Probabilities……………………………………..…..209

CHAPTER 1

SIX SYSTEMATIC STEPS FOR SOLVING GENETICS PROBLEMS

Solving genetics problems is not very difficult if you are systematic in your approach. The trick is to recognize the inheritance pattern that each problem contains. Only a few basic patterns exist, so once you have learned to spot them, you will find that you can recognize most any problem. That's why it is important to do many genetics problems to learn genetics; the more problems you do, the more practice you will get at recognizing the patterns.

Step A. Always use standard designations for the generations and lay out the problem as a simple diagram.

> When you cross two organisms, they are called the **parents** or the **P** generation. Their offspring are called the **first filial generation** or F_1. ("Filial" comes from a Latin word for son or daughter.) If you cross two F_1 individuals, their offspring are called the **second filial generation** or F_2. These terms are always used in genetics problems; understand the terms and use them whenever working a problem.
>
> Many problems are stated in paragraph form, often covering up the basic pattern with words. You will find it easier to do most problems if you change the information into a simple diagram of the crosses, making the patterns much easier to recognize.

Solved Problem 1-1. Although flies usually have a pair of wings, geneticists have found some wingless strains. When one of these wingless flies was crossed to a normal fly with wings, all of their offspring had wings. When some of these offspring were crossed to each other, they produced 428 offspring, of which 320 had normal wings and 108 were wingless. Summarize this cross in a diagram, using the correct terms for each generation.

Answer. The three generations can be summarized in this way:

P: winged X wingless

F_1:   all winged

F_2:   320 winged, 108 wingless

PRACTICE PROBLEMS (Answers to all practice problems begin on p. 152)

1-2. A white bison was born in a rancher's herd. When it matured, the rancher mated it to a normal bison on three occasions, and each time a normally colored bison was born.

When those new bison grew up, two of them were mated to each other. All of their calves were brown except one, which was albino like its grandmother. Summarize this cross in a diagram, using the correct terms for each generation.

1-3. Silk worm blood (or "hemolymph") is either deep yellow or white. The caterpillar may be plain or heavily marked with stripes and spots. A moth with yellow hemolymph that had been a plain caterpillar was crossed to a moth with white hemolymph that had been a marked caterpillar. Among the offspring were 65 yellow, marked; 56 yellow, plain; 61 white, marked; and 59 white, plain. Diagram this cross using the correct terms for each generation.

1-4. In a certain locality almost all lacewings (an insect of the order Neuroptera) had golden eyes, but a small number had green eyes. When a golden-eyed lacewing was crossed to a green-eyed lacewing, some golden-eyed and some green-eyed lacewings were found among the offspring. Diagram this cross.

*1-5. Mastiff dogs may suffer from an ocular disease called "progressive retinal atrophy," or PRA. Two different crosses of affected dogs to dogs without the disease were among those crosses used to characterize the PRA allele. In a first cross, an affected male was crossed to an unaffected female to produce a litter of five dogs: two unaffected males, and one affected, and one unaffected female, and one affected. In a second cross, an unaffected male was crossed to an affected female, and the litter contained four males, one of which was affected and two females, one affected and one not. Diagram these crosses. (Kijas *et al.,* 2003)

Step B. Write a key for the symbols you are using for the allelic variants of each locus.

In order for genetics **crosses** to be done, gene differences have to be identified. Little would be learned by crossing pure-breeding red-flowered plants generation after generation except that there were no gene differences. An important insight of Mendel's was that the genes act like particles that are passed on from generation to generation, and those particles may occur in more than one form. The particle is called a **locus** and each different form is called an **allele** of that locus.

The simplest symbol system designates **dominant** alleles with an upper case letter and **recessive** alleles with the same letter in lower case. For example, in humans the recessive allele for sickle cell anemia can be symbolized by *s*, and the dominant allele can be symbolized by *S*. The two alleles are both at the S locus or the sickle cell locus. The symbols only tell us which allele is dominant and which recessive, not which one is "normal" or the most frequent. For example, the dominant mutation "Bar eye" in fruit flies reduces the number of facets in the compound eye. Flies with normally shaped eyes have the recessive allele of this locus, so Bar eye is *B* and normal eye is *b*. (Scientists studying particular cases may use more complex symbols, as, for example, when there was no dominance. We will need such symbolism only rarely in this book.)

When working with **diploid** organisms, each individual will contain two alleles for each locus. If the individual is **homozygous**, we symbolize it with two of the same allele symbol, but if this individual is **heterozygous**, we use two different allele symbols. Gametes, which are **haploid,** are designated with single allele symbols for each locus. If more than one locus is being symbolized in one individual, separate the symbols of different loci with a space.

Solved Problem 1-6.   Mendel studied seven different loci in peas, with two alleles possible at each locus. Let the symbol for tall be *T* and short be *t* and let the symbol for green pods be *G* and yellow pods be *g*, then write the symbols for each of the following:

A) a homozygous tall plant   D) a plant heterozygous at both loci
B) a plant heterozygous for pod color   E) a gamete with the tall allele
C) a short plant with yellow pods   F) a gamete with short and green alleles

Answer. A) *TT* B) *Gg* C) *tt gg* D) *Tt Gg* E) *T* F) *t G*

Solved Problem 1-7. *Drosophila* usually have round eyes, but a recessive mutation results in eyeless flies. They usually have straight bristles on their backs, but a recessive mutation results in crooked bristles that are twisted like little corkscrews. Finally, a dominant mutation changes the normal brick red eyes to a purple color called "plum." Symbolize a fly heterozygous at all three loci.

Answer. You will need to decide what letter you want to use for each locus. Suppose you use these symbols:

$$E = \text{round}; \quad e = \text{eyeless}$$
$$C = \text{straight}; \quad c = \text{crooked}$$
$$P = \text{plum}; \quad p = \text{red}$$

Then the fly in question would be *Ee Cc Pp*.

PRACTICE PROBLEMS

1-8. Huntington's disease is a human malady caused by a dominant allele that is not expressed until a person reaches middle adulthood. Cystic fibrosis is caused by a recessive allele at a different locus. Choose the letters you want to use and then write the symbols for each of the following individuals.

  A) A young person who is destined to have Huntington's disease later in life.
  B) A person who does not have alleles for Huntington's disease or cystic fibrosis.
  C) A person who has cystic fibrosis, but will not develop Huntington's disease.

1-9. A homozygous fly with an ebony body color (recessive allele) was crossed to a fly homozygous for the wild-type straw-colored body (dominant), and all their offspring were heterozygous for this body color locus. Write symbols for each parent and for their offspring.

1-10. The golden-eyed lacewings in the P generation of Problem 1-4 are heterozygous at a locus for eye color, and the green-eyed parents are homozygous for the recessive allele. Choose a letter and write symbols for the golden-eyed and green-eyed parents.

*1-11. Flower color in periwinkles involves many loci. One locus has the *R* and *r* alleles, a second locus has *W* and *w*, and a third has the *E* and *e* alleles. Write each of these genotypes. (Sreevalli, Kulkarni, and Baskaran, 2002)

  A) Heterozygous for all three loci.
  B) Homozygous for *e* and *R*. Heterozygous for the remaining locus.
  C) Homozygous for the three recessives.

Step C.   Determine the genotypes of the parents of each cross.

Geneticists deal with two general features of organisms.   The **genotype** is the collection of genes that an organism has, while the **phenotype** is a description of the characteristics of an organism.   Most simple genetics problems are of two general kinds.   One, covered in this section, requires you to find the genotypes or phenotypes of parents, while the other, requires you to find the genotypes or phenotypes of offspring.   Three kinds of evidence can be used to figure out the genotypes of the parents.

First Evidence: Are the parents from true-breeding lines?

If a line results in the same kind of offspring generation after generation, it is a true-breeding line. **Such a line must be homozygous.** Sometimes a problem will simply say that the lines used are true-breeding. Sometimes other data allow you to figure that out.

Solved Problem 1-12.   Which of the following individuals are probably homozygous?

A) A rat from a true-breeding strain of laboratory animals.
B) Two parents with brown eyes that produce a child with blue eyes.
C) A snapdragon with purple flowers from a strain that had consistently produced purple flowers for 14 generations.

Answer.

A) If the rat is from a true-breeding strain that means it is, by definition, homozygous.
B) The parents are not true-breeding, so at least one of them is not homozygous. In this case, probably neither is. To produce a child different from themselves, each must have contributed a recessive allele but must also have a dominant allele.
C) "True-breeding" means that the strain produces the same kind of individuals in every generation, so this plant is from a true-breeding strain and thus is homozygous.

PRACTICE PROBLEMS

1-13.   Jake made a deal with his father-in-law, Lavern. They divided up Lavern's flock so that Jake got all the spotted goats and the black sheep, while Lavern got the solid colored goats and the white sheep. They agreed that whenever a spotted goat or black sheep appeared in Lavern's flock, he would give it to Jake; and whenever a solid goat or a white sheep appeared in Jake's flock, he would give it to Lavern. As time passed, Lavern had to keep giving some of the offspring from his flock to Jake, but Jake didn't have to give any of his flock to Lavern. Explain using the terminology of this section.

1-14. Sometimes pairs of wild foxes produce one or more silver-backed pups. Two such pups raised in captivity always have silver-backed pups when mated to each other. Which individuals in this example are homozygous?

*1-15. In an effort to find more gene loci in the mosquito *Anopheles gambiae* (principal vector of malaria in sub-Saharan Africa), mosquitoes were irradiated with radioactive cobalt and then inbred. One particular variant that arose was *homochromy I* (homo-chromy-one), a variant in body color. Twenty-five *homochromy I* individuals were obtained, all of which established pure lines when mated with one another. Is *homochromy I* due to a dominant or recessive allele? (Benedict *et al.*, 2003)

Second Evidence: Can the parents' genotypes be reliably deduced from their phenotypes?

A dominant allele is one that is expressed whether it is homozygous or heterozygous, while a recessive allele is only expressed when homozygous. If we know which allele is dominant and which recessive, we can know the genotype of recessive parents with certainty. If we know nothing about an individual except that it has the dominant phenotype, then its genotype is ambiguous—it could be either homozygous or heterozygous, and other information will be needed to know which it is for certain.

Solved Problem 1-16. Look at the cross described in Problem 1-1. Suppose that lack of wings is recessive and is symbolized with a lower case *w* and the allele for having wings is *W*. On the basis of that information alone, write the genotypes of all the individuals in the parental, F_1 and F_2 generations that you know for certain.

Answer. All the wingless flies are unambiguously *ww*. On the basis of only our knowledge of the parental phenotypes, we are not able to tell with certainty what the genotypes of winged flies might be. But since the winged parent must have at least one *W* allele, it can be symbolized *W_*, with the blank for the unknown allele.

P: winged (*W_*) X   wingless (*ww*)

Solved Problem 1-17. Smooth fox terriers sometimes have congenital myasthenia gravis, a muscle disease caused by a recessive allele that produces a defect in the nerve-muscle synapse when homozygous. Such dogs almost always die by 6 to 9 weeks of age. Can we ever know for certain the genotype of parent fox terriers by looking at their phenotype?

Answer. No, we cannot. The recessive homozygotes have myasthenia gravis and die as puppies. So they can never be parents. Parents will not have myasthenia gravis, so they could be homozygous or heterozygous.

PRACTICE PROBLEMS

1-18. Refer to Problem 1-13. Assuming those coat colors and patterns are due to single gene differences, can we know any of the genotypes with certainty?

1-19. Peroneal muscular atrophy afflicts humans, beginning when the individual is 10 to 20 years old. A person develops peroneal muscular atrophy only if at least one of his or her parents had it. Assuming that a single gene difference is at issue here, can you tell whether the gene is caused by a dominant or recessive allele based on these data?

*1-20. A bell pepper strain called "flaccid" had droopy leaves and stems. When two flaccid plants were crossed, all the offspring were flaccid. When two plants of a strain with normal leaves (called strain KRG) were crossed, all the offspring were normal. When KRG was crossed to flaccid, all the offspring were normal. Assuming that alleles at a single locus make the difference between normal and flaccid, which strains have genotypes we can know for sure? (Bosland, 2002)

Third Evidence: Do the phenotypes of the parents' offspring provide any information?

A diploid organism provides one allele of each locus to its offspring. The offspring receives one allele from each parent. Sometimes this makes it possible to deduce the genotype of a parent by looking at the offspring. An organism that has a recessive phenotype must be homozygous recessive and must have received one recessive allele from each parent. So each parent must have at least one recessive allele to give to the offspring. If a parent has the dominant phenotype and there are offspring with the recessive phenotype, that parent must be heterozygous.

> **Solved Problem 1-21.** A man's wife had an albino child. Albinism, a lack of melanin pigment, is caused by a recessive allele of a single gene. "Neither you nor I is albino," said the man. "Therefore, I am not the father of this child." Is the man correct?
>
> **Answer.** Let's symbolize the recessive albino allele as *a* and the dominant allele that produces melanin as *A*. Let's diagram the cross and write what we know for sure.
>
> P: melanin (??) X melanin (??)
>
> F$_1$: one albino child (*aa*) (we know this genotype because the
>   phenotype is the recessive one)
>
> Because the parents are not albino, they could be either *AA* or *Aa*. A way of symbolizing that is to write their genotypes as *A_* , where the blank could be either *A* or *a*.
>
> P: melanin (*A_*) X melanin (*A_*)

F$_1$: one albino child (*aa*)

The albino child must have received an albino allele from each parent in order to be *aa*, so that allows us to fill in the blanks in the parents' genotypes.

P: melanin (*Aa*) X melanin (*Aa*)

F$_1$: one albino child (*aa*)

So the man could be the father of an albino child even if neither he nor any of his relatives is albino.

Solved Problem 1-22.   Refer to Problems 1-13 and 1-18. If Lavern knew anything about genetics, he might prefer to kill and eat the ewes (female sheep) in his flock that gave birth to black sheep or perhaps, less drastically, sell them to someone else. Why would these be possible strategies?

Answer.   White ewes have the dominant phenotype. If they give birth to black sheep (*ww*), they must be *Ww*, since the offspring get a *w* from each parent. If he eliminates the heterozygotes, he won't have to keep giving sheep away to his son-in-law.

PRACTICE PROBLEMS

1-23.   What is certain about the genotypes of the winged flies in Problem 1-16?

1-24.   Farmer O'Sullivan has a prize boar named Honey that has won many honors at the State Fair. Several other farmers want to mate Honey to their sows so they can have prize pigs too. But Farmer O'Sullivan knows that some of Honey's litter mates had been runts and that the failure to grow properly was due to a recessive gene. What does this tell us about Honey's possible genotypes?

1-25.   What cross could Farmer O'Sullivan do to find out for certain what Honey's genotype is?

1-26.   *Paramecium tetraurelia* usually survive concentrations of potassium in the culture fluid as high as 30 millimolar. However, recessive alleles of two different loci make the protists more sensitive. They die when potassium levels reach 20 millimolar. The sensitive allele of one locus is *k1*, and the sensitive allele of the other locus is *k2*. When either allele is homozygous, the protozoa are potassium-sensitive. There must be a dominant allele (*K1* and *K2*) at each locus for the protozoa to show normal resistance. Which of the following genotypes exhibit potassium sensitivity and which are resistant?

A) *k1k1 K2k2* B) *K1K1 K2K2* C) *K1k1 K2k2* D) *k1k1 K2K2* E)*K1K1 k2k2*

*1-27.   With the completion of the human genome project, people are finding genes in new ways. Now they can compare a sequence of DNA from the genome project to a particular gene they are looking for. In this way, a gene was found in humans for a receptor in the tongue involved in sensing sweet tastes. They called this locus "tas" for taste receptor and found it to be the same locus as "Sac", a sweet responsiveness locus in mice that has nearly the same structure and is located at nearly the same site on its chromosome as the human gene. In searching the data banks of DNA sequences, the investigators found two taster alleles and two non-taster alleles of "tas." Suppose they wanted to know which was dominant and which recessive, a taster or non-taster. Suppose further that they examined a large number of people in a population and identified each one as taster or non-taster.

The organisms are people in this case, so you can't just do various matings of your choice. You have to take the matings the subjects give you. Humans tend to produce small numbers of offspring too. If they wanted to know whether taster or non-taster is dominant, what mating would give an unambiguous answer? (Max *et al.*, 2001)

Step D. Indicate the possible kinds of gametes formed by each of the parents.

> Sexual reproduction involves the alternation of **fertilization**, which produces **zygotes**, and **meiosis**, which precedes gamete formation. When laying out a problem, think about the sequence of fertilization - gamete formation - fertilization - gamete formation. Then by systematically filling in what you know and using that information to deduce what you don't know, you can solve most any genetics problem. In this section we will work out the gamete types.

Monohybrid cross? Apply the Principle of Segregation.

In a **monohybrid cross** we focus on a single locus. When there are just two possible alleles, gamete formation is simple. Each gamete will receive one allele. If the parent is homozygous, all its gametes will be alike. If the parent is heterozygous, each gamete will receive only one of the two possible alleles. This is what is meant by the **Principle of Segregation**.

> **Solved Problem 1-28.**   What are all the types of gametes formed by individuals of each of the following kinds? What proportions of every type of gamete will each individual produce?
>
> A) *QQ*   B) *Ww*   C) *mm*
>
> **Answer.**   Remember, each gamete will receive only one allele, and if there are two possible alleles, half the gametes will get each type.
>
> A) All the gametes of *QQ* will contain allele *Q*.
> B) Half the gametes of *Ww* will contain allele *W* and half will contain *w*.
> C) All the gametes of *mm* will contain allele *m*.

PRACTICE PROBLEMS

1-29.   What possible alleles will be found in gametes produced by chickens from a true-breeding white-feathered strain?

1-30.   Refer to Problem 1-19. What possible alleles will be found in gametes produced by each parent?

1-31.   If a geneticist discovered an individual heterozygous for an allele that killed any sperm that had that allele, could she maintain a stock of organisms that had this gene?

*1-32.   Look at Problem 1-11. What gametes could be produced by homozygous *E*, homozygous *e* and the heterozygote?

Dihybrid cross?   Apply both the Principle of Segregation and the Principle of Independent Assortment.

If two loci are being studied simultaneously (a **dihybrid cross**), then the problem might seem a little more complicated, but it really isn't. If you have two loci, you can treat each like a monohybrid cross. Even if there are many more loci than two, you can still do this and have a simpler problem to solve. Each gamete will only get one allele from each locus. In other words, all the loci are acting according to the Principle of Segregation.

In every gamete, a particular allele of one locus can be combined with either of the alleles of the other locus. That is what the **Principle of Independent Assortment** means.

The important thing is to be sure that you have accounted for all of the possible combinations of alleles when you figure out the gametes. Two accounting methods make this rather easy.

The branch method. Write down the genotype of the first locus and then make a branch showing the kinds of gametes expected from that locus. Suppose for example, that we were working with an organism heterozygous for two loci, *Aa Bb*.

Write a branch diagram for the first locus like this:

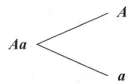

The *A* allele and the *a* allele can now each be combined with either of the choices for the next locus, so write a second branch diagram next to the end of each of the first branches. Follow the branches on this diagram to get gamete types.

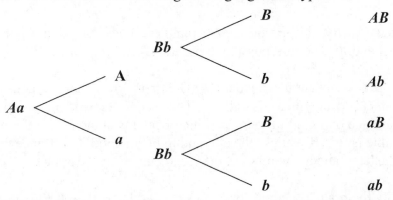

The alternating list method. If you know how many gametes you should end up with, then it is easier to be sure you have listed all the possible types. This second method for accounting for all the gametes is based on the observation that the number of different kinds of gametes will be equal to 2^n where *n* is the number of loci that are heterozygous.

Again, suppose the organism is the double heterozygote, *Aa Bb*. Because there are two heterozygous loci, there will be 2^2, or four, different kinds of gametes. Of the four gametes, two will have an *A* allele and two will have an *a* allele. List these like this:

A

A

a

a

Of the gametes with an *A* allele, half will have *B* and half will have *b*. Likewise, half of those with the *a* allele will have *B* and half will have *b*. So now write a second column on your list like this.

AB

Ab

aB

ab

If you did it correctly, the alleles of the last locus on your list will alternate every other one, as in this case.

Although we will be concentrating on crosses with only two loci, both the branch and list method will work no matter how many loci you are working with. If you want to be sure you have the idea for the dihybrid cases, try some of the later problems in this section to see if you can do it with a larger number of loci.

Solved Problem 1-33. In each of the following cases, use either the branch or the list method to determine the kinds and proportions of gametes that the individual will produce.

A) *Mm Nn*   B) *AA bb*   C) *RR Tt*

Answer. A) This is exactly like the previous example except that *M* and *N* have been substituted for *A* and *B*. The four possible types are *MN, Mn, mN* and *mn*.

B) Remember that the two loci act independently. The first locus is homozygous and can only produce gametes containing an *A* allele. The second locus is homozygous, so it too can only produce one kind of gamete, which contains allele *b*. So the only kind of gamete possible is *Ab*. If you use the alternating list method, n = 0 since there are no heterozygous alleles, and the number of different gametes is 2^0, or one.

C) To do this one by the branch method, first consider the locus that is *RR*. It will make only one kind of gamete, containing *R*.

RR ————— *R*

Then the second locus can form either *T* or *t* gametes, so

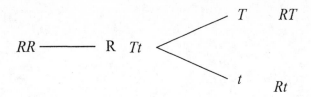

By the list method, first determine the number of possible gamete types, which is 2^1, or two since there is only one heterozygous locus. There will be two kinds of gametes, and all will contain *R*.

> *R*
> *R*

Then half of those will contain *T* and half *t*, so:

> *RT*
> *Rt*

Solved Problem 1-34. Tay-Sachs disease is a serious human recessive disorder that affects newborns who are homozygous for the deleterious allele *t*. People with a single normal allele *T* do not have Tay-Sachs disease. However, they can be detected as heterozygotes for this locus by biochemical screening. A dominant gene *P* produces a characteristic called "polydactyly" in which humans have one or more extra fingers or toes. A person with polydactyly who had one parent with normal fingers and toes is found to be heterozygous for Tay-Sachs disease. What are all the kinds of gametes this person can produce?

Answer. We know the person is heterozygous *Tt* because the problem says so. A person with the dominant trait polydactyly is either *PP* or *Pp*. Because the person has one normal parent (*pp*), that parent would have produced only *p* gametes, so the person under consideration must be *Pp*. The question thus breaks down to "What are all the kinds of gametes that can be produced by a person who is *Pp Tt*?" The answer is *PT, Pt, pT* and *pt*.

Solved Problem 1-35. Write all the kinds of gametes that can be produced by a person with this genotype: *Aa BB cc Dd Ee ff gg HH*.

Answer. At first this looks rather daunting, but take it one step at a time. Let's use the alternating list method, although either method would work. There are eight loci in this problem, but only three are heterozygous. So the total number of different gametes that can be produced is $2^3 = 8$. Half will contain allele *A* and half will contain allele *a*. So start the alternating list this way:

> *A*　　　　　　*a*
> *A*　　　　　　*a*
> *A*　　　　　　*a*
> *A*　　　　　　*a*

Now all those gametes will have a *B* allele and a *c* allele because these two loci are homozygous.

A B c	*aBc*
A B c	*aBc*
A B c	*aBc*
A B c	*aBc*

The next two loci are heterozygous, and in each instance an upper case allele will end up in half the gametes and a lower case allele in the other half. Do the alternating list for each.

A B c D E	*aBcDE*
A B c D e	*aBcDe*
A B c d E	*aBcdE*
A B c d e	*aBcde*

The last three alleles are all homozygous, so the same allele will end up in each gamete. The final list will therefore look like this:

A B c D E f g H	*a B c D E f g H*
A B c D e f g H	*a B c D e f g H*
A B c d E f g H	*a B c d E f g H*
A B c d e f g H	*a B c d e f g H*

PRACTICE PROBLEMS

1-36. Farmer Vandermeer had two hogs named Olga and Hank who were normal healthy hogs, but whenever he crossed them, some of the offspring had cleft palates. Assuming cleft palate has a genetic basis, what are all the kinds of alleles in gametes that Olga and Hank can produce?

1-37. If a *Drosophila* is heterozygous for recessive alleles *e* (ebony), *f* (forked), *t* (temperature-sensitive) and *v* (vestigial), what kinds of allele combinations can that fly produce in its gametes?

1-38. If an organism were heterozygous for 10 loci, how many different allele combinations could the organism produce in its gametes?

*1-39. As described in Problem 1-11, the flowers of the periwinkle (*Catharanthus roseus*) come in many different colors due to many gene interactions. The corolla is the outer part of the flower, and the eye is the center part. Investigators crossed a plant homozygous for four different loci *ee RR ww OO* to another plant also homozygous, *EE* *rr WW oo*. Flowers of the first plant have orange-red corollas and white eyes. Flowers of the second plant are all white. The flowers of the F_1 have a pink corolla and a red eye. (Sreevalli, Kulkarni, and Baskaran, 2002)

A) What are all the gamete types that each parent and the F_1 can produce?

When two F_1 were crossed, progeny included these three genotypes. What gametes in what proportion can each of the following F_2 plants produce?

B) *Ee RR Ww oo*   C) *EE RR WW oo*   D) *ee rr Ww oo*

Show your work using the alternating list method for the parents and F_1 in (A). Use the branch method for (B), (C), and (D). (Srevalli, Kulkarni, and Baskaran, 2002)

Step E. Set up a Punnett square.

Fertilization follows gamete production. At fertilization, the gamete of one parent combines with that of the other. There are many potential combinations. One way to identify these possible combinations is with a **Punnett square**. This is a grid with the gametes of one parent written across the top and the gametes of the other written along the left side.

Solved Problem 1-40. Suppose we crossed two organisms, each of which had the genotype *Bb*. Set up a Punnett square for that cross. What if one parent were *Bb* and the other were *bb*?

Answer. Set up a simple diagram of the cross to show what kinds of gametes each parent will produce:

P: *Bb*   X   *Bb*

Gametes   *B* and *b*   *B* and *b*.

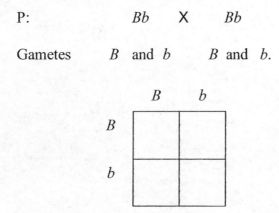

If the parents were *Bb* and *bb*, the gametes will be a little different, but the approach will be the same.

P: *Bb*   X   *bb*

Gametes   *B* and *b*   all *b*

(Just one kind of gamete so just one row.)

Solved Problem 1-41. Prepare a Punnett square for a cross of two parents who are *Aa Bb*.

Answer. Do a simple diagram of the cross and then set up the Punnett square.

P: *Aa Bb* X *Aa Bb*

Gametes *AB Ab aB ab*   *AB Ab aB ab*

	AB	*Ab*	*aB*	*ab*
AB				
Ab				
aB				
ab				

PRACTICE PROBLEMS

1-42. Suppose we cross an organism of genotype *MM NN* with one that is *mm nn*. Prepare a Punnett square for this cross.

1-43. Set up a Punnett square for a cross of two flies heterozygous for *B*, *F* and *G*.

1-44. If a mouse was heterozygous at 14 different loci, and its mate was heterozygous for just three of those loci, what would be the dimensions of a Punnett square used to study this cross?

*1-45. Bell pepper plants from the flaccid strain have droopy stems and leaves, while plants from the KSG strain all have erect stems and leaves. (See Problem 1-20). All the offspring of a cross of flaccid X KSG are normal. Some of the bell peppers from this first cross were self-fertilized, while others were crossed to either the original KSG or flaccid strains. Thus, four crosses were done. Assuming that flaccid is due to a recessive allele we'll call *f* and normal occurs due to a dominant allele *F*, set up a Punnett square for each cross. (Bosland, 2002)

Step F. Fill in the Punnett square and determine the genotypic and phenotypic ratios of the offspring.

With the Punnett square set up, it is now a simple matter to fill in all the intersections with the alleles from both gametes that form that intersection. In this way you can be sure of having represented all possible gamete combinations. Then, once the genotypes of the new zygotes are determined, you can figure out the phenotypes of each zygote type to complete the problem.

Solved Problem 1-46.   Fill in all the intersections on the Punnett squares in Problems 1-40 and 1-41. Suppose that *b* was a recessive allele that caused abnormal flying pattern in pigeons, and *B* was a dominant allele that resulted in normal flight. After you figure out what genotype is to be found in each intersection, put in each associated phenotype.

Answer.

	B	*b*
B	*BB* *Normal*	*Bb* *Normal*
B	*Bb* *Normal*	*bb* *Abnormal*

	B	*b*
B	*Bb* *Normal*	*bb* *Abnormal*

When Punnett squares get a little complicated, as in Problem 1-41, follow a systematic pattern for filling in the intersections so you won't get confused.

1) Fill in just one allele at a time, and keep all the alleles of the same locus together. So in the example below, note that all the *A*s are together in any one intersection and all the *B*s are together. Fill in the *A*s first, then do the *B*s.

2) When there are two different alleles at an intersection, put the upper case letter first. Notice that there are no genotypes written *aA* or *bB*.

	AB	*Ab*	*aB*	*ab*
AB	*AA BB*	*AA Bb*	*Aa BB*	*Aa Bb*
Ab	*AA Bb*	*AA bb*	*Aa Bb*	*Aa bb*
aB	*Aa BB*	*Aa Bb*	*aa BB*	*Aa Bb*
ab	*Aa Bb*	*Aa bb*	*aa Bb*	*Aa bb*

PRACTICE PROBLEMS

1-47. Fill in all the intersections on the Punnett square of Problem 1-42.

1-48. Fill in all the intersections on the Punnett square of Problem 1-43.

1-49. How many intersections will the Punnett square of Problem 1-44 have?

***1-50.** Read Problem 1-45 again, and then fill in the four Punnett squares and indicate the phenotype of the genotype in each intersection. (Bosland, 2002)

IN SUMMARY: THE SIX (OR SEVEN) SYSTEMATIC STEPS

Step A. Make a Diagram. Use standard designations for the generations and lay out the problem in a simple diagram.

Step B. Choose Symbols. Make a key to your symbols for the alleles.

Step C. Discover Parental Information. Determine the genotypes of the parents in each cross. To do so, use this evidence:

- Are the parents from true-breeding lines?
- Can their genotypes be deduced from their phenotypes?
- Do the phenotypes of their offspring provide any information?

Step D. Decide the Parent Gamete Types. Indicate the possible kinds of gametes formed by each of the parents.

If a monohybrid cross, use the Law of Segregation.
If a dihybrid cross, use the Law of Independent Assortment.

Step E. Set Up a Punnett Square. Set up a Punnett square to show all the zygote types.

Step F. Fill in the Punnett Square. Fill in the Punnett square and determine the genotypic and phenotypic ratios of the offspring.

[Note: **Some people add a Step G. See Chapter 3. Use the Product Rule.** Use the Product Rule of Probability as a shortcut to determining predicted ratios, when possible. You will find it useful if you do many complex problems. The material on pages 29-34 will help you if you need to learn Step G.]

CHAPTER 2

GOING BEYOND THE SIX STEPS

The combination of the six steps that you use depends on the kind of problem you have to solve.

In general, there are two kinds of genetics problems, those that tell you about the parents and ask for conclusions about the offspring and those that tell you about the offspring and ask for conclusions about the parents. These problems require slightly different approaches.

Given parental data, make conclusions about the offspring.

This is the most direct kind of problem. Steps D through F are most helpful, although you will still benefit from simplifying the problem with a diagram, and you will need to define the gene symbols to be used.

Solved Problem 2-1. The *p* allele of the German cockroach produces a fine mottling all over the body. This condition is called "peppery" and is recessive to the wild type allele, which produces solid coloring. What are the expected genotype and phenotype ratios among the offspring in each of these crosses?
   A) A cross between two peppery cockroaches.
   B) A cross between two heterozygous cockroaches.
   C) A cross between a peppery cockroach and a heterozygote.

Answer. Diagram each cross, put down all that you know and then construct Punnett squares to get the answer.

A)   P: Peppery   X   Peppery
                    pp   *pp*

      Gametes   all *p*   all *p*

      Punnett square:

                            p
                    ┌──────────────┐
               p │   *pp*   │
                    │   peppery   │
                    └──────────────┘

      Genotypes: all *pp*. Phenotypes: all peppery.

B)   P: Solid   X   Solid
                              Pp   *Pp*

Gametes: 1/2 *P*   1/2 *P*
                    1/2 *p*   1/2 *p*

Punnett square:

	P	*p*
P	*PP* Solid	*Pp* Solid
p	*Pp* Solid	*pp* Peppery

Genotypes: 1:2:1.   Phenotypes: 3 solid:1 peppery.

C)   P: Peppery X   Heterozygote (Solid color)
                            pp   *Pp*

Gametes   all *p*   1/2 *P*
                                        1/2 *p*

Punnett square:

	P	*p*
p	*Pp* Solid	*pp* Peppery

Genotypes: 1:1.   Phenotypes: 1 solid:1 peppery.

Solved Problem 2-2. A plant had the genotype *Ff Gg Hh*. If two such plants were crossed, 1) what proportion would show the dominant phenotype for *F*? 2) what proportion would show the recessive phenotype for all three genes?

Answer. Since there are three heterozygous loci, there will be eight different possible gametes:

FGH   *FGh*   *FgH*   *Fgh*

fGH   *fGh*   *fgH*   *fgh*

If you do these problems with a Punnett square, there will be 8 X 8, or 64, boxes to fill in, but you don't have to do that given what the problem asks. You can use the Product Rule of Probability. This is covered in detail in Chapter 3 (on pages 29).

1. This question is only about F and f. The loci are independently assorting, so you don't have to consider G and H. In a cross of Ff X Ff, 1/4 will be FF and 1/2 will be Ff. So 3/4 will have the dominant phenotype.

2. In a cross of Ff X Ff, ff will occur 1/4 of the time. Likewise, in a cross of Gg X Gg, gg will occur 1/4 of the time, and in a cross of Hh X Hh, hh will occur 1/4 of the time. So the genotype $ff\,gg\,hh$ will occur 1/4 X 1/4 X 1/4, or 1/64 of the time.

PRACTICE PROBLEMS

2-3. Look at Problem 2-2. If one of those triple heterozygotes were crossed to a plant homozygous for all three recessive alleles, what would be the proportions of each genotype among the offspring?

2-4. In humans, polydactyly is caused by a dominant allele and results in extra digits. Phenylketonuria is recessive and is a condition caused by a disorder in the metabolism of phenylalanine. Unless given a special diet as infants, people with phenylketonuria may have varying levels of mental retardation. A man who has neither condition but whose father has phenylketonuria and a woman with no phenylketonuria allele but with polydactyly like her father (assume she is heterozygous for polydactyly) are considering having children. What is the probability that they will have a child with both conditions? What about with one of these conditions?

2-5. Recessive alleles of two different loci in mice give the very same phenotype. One is light ear (l) and the other is pale ear (p). Each results in ears that are lighter than normal, and each has a dominant allele, L or P, that results in the normal dark ear. The loci are independently assorting. Investigators wanted to create a double mutant strain $ll\,pp$. After a series of crosses, they had a group of light-eared mice, some of which were $ll\,PP$, some $LL\,pp$ and some $ll\,pp$. But they all looked alike. They had the original strains of each single mutant ($LL\,pp$ and $ll\,PP$) as well as the wild type $LL\,PP$. What crosses can they do to distinguish between the single and double homozygous recessives?

*2-6. The beetle *Rhysopertha dominica* is a harmful agricultural pest in Australia. Two loci have been found which have recessive alleles that confer resistance to the insecticide phosphine. Call the recessive alleles $p5$ and $p6$. When homozygous, allele $p5$ raises the level of resistance to phosphine 50-fold, and when $p6$ is homozygous, it raises resistance 12.5-fold. When both alleles are homozygous, phosphine resistance is increased 250-fold.

For each of these crosses, predict the resistance of the progeny in the F_1 and F_2 generations.

A) $p6\,p6$ X $P6\,P6$ B) $p5p5$ X $P5P5$ C) $p5\,p5\;P6\,P6$ X $P5\,P5\;p6\,p6$

(Schlipalius *et al.,* 2002)

Given data about offspring, make conclusions about parents.

> Here you will work backward. After you have done a number of problems of the first kind, you will see that there are just a few possible ratios that you get among the offspring. Recognizing these ratios is very important because when you know the offspring ratios, that is almost always useful in making conclusions about the parents.

If the offspring ratio is:	Then we know this about the parents:	The cross is called:
3:1	Both parents are heterozygous at one locus. One allele is dominant.	Monohybrid Cross
1:1	One parent is homozygous recessive. One parent is heterozygous.	Test Cross
9:3:3:1	Each parent is heterozygous at two loci. One allele is dominant at each locus.	Dihybrid Cross
1:1:1:1	One parent is a double heterozygote. One parent is homozygous recessive for both loci.	Two-Point Test Cross
	-OR-	
1:1:1:1	One parent is heterozygous at the first locus and homozygous at the second. One parent is heterozygous at the second locus and recessive at the first (e.g., *Bb tt* X *bb Tt*).	

Solved Problem 2-7. A tropical fish fancier produced an albino zebra fish. It lacked the blue stripes characteristic of the normal zebra fish, and it had pink eyes. When he crossed a normal zebra fish to the albino, all the offspring were normal. When he crossed two of the normal F_1 offspring, 1/4 were albino and 3/4 were normal. When an albino was crossed to the normal F_1 offspring, half were albino and half were normal. What is the genotype of each fish mentioned in this problem?

Answer. Look for a cross that gives familiar results. There is a 3:1 ratio among the progeny of the F_1. This is what we expect when two individuals are crossed that are both heterozygous at a single locus. So that cross would be:

Normal (*Aa*) X Normal (*Aa*)

When albino is crossed to normal, the offspring are normal, suggesting that albino is the recessive allele. So the original parental fish were a normal *AA* and an albino *aa*.

The last cross of albino to the F_1 normal confirms these previous identifications. The 1:1 ratio in that cross is characteristic of a test cross in which one fish (the normal) is *Aa* and the other fish (the albino) is *aa*.

Solved Problem 2-8. A mutation in chickens results in extra bones in the wings and feet, a shortened beak, and an inability to hatch without help because they are unable to peck their way out of the shell. The mutation is a recessive allele called *d* for "diplopodia." A cross was done involving diplopodia and another locus that controls feather color called "dominant white." The *I* allele results in white feathers, but *i* results in colored feathers. In a particular cross, the offspring included 180 white chickens; 68 with colored feathers and diplopodia; 166 with colored feathers; and 53 with white feathers and diplopodia. Figure out the genotypes and phenotypes of the parents.

Answer. We can write down quite a bit that is known about the offspring without considering the ratios at all:

180 white (no diplopodia)
68 color and diplopodia
166 color (no diplopodia)
53 white and diplopodia

Remember that we know the dominance relationships of these genes. White is dominant to color and normal is dominant to diplopodia so:

180 white (no diplopodia)	*I_ D_*
68 color and diplopodia	*ii dd*
166 color (no diplopodia)	*ii D_*
53 white and diplopodia	*I_ dd*

Now consider each locus separately to simplify the problem. (Notice how often simplification comes from the approach of considering one locus separately.)

The feather color phenotypes occur in a ratio of about 1:1 (233 white:234 color). That is characteristic of a test cross, *Ii* X *ii*.
The diplopodia phenotypes occur in a ratio of about 3:1 (346:121). That is characteristic of a cross of two heterozygotes where one allele is dominant, or *Dd* X *Dd*.

So the original cross was probably *Ii Dd* X *ii Dd.*

PRACTICE PROBLEMS.

2-9.   A cross in chickens involved the diplopodia gene (see Problem 2-8) and "naked neck" (*n*), a recessive mutation at another locus that results in loss of feathers on the neck. The offspring included 85 normal chickens; 10 with naked neck and diplopodia; 26 with naked neck; and 30 with normal neck and diplopodia.   What were the genotypes and phenotypes of the parents?

2-10.   Horses can have big white spots (a recessive allele *b*) or little white spots (a dominant allele *B*).   They can also have weak hooves (*W*) or strong hooves (*w*), and normal nostrils (*F*) or flaring nostrils (*f*).   A woman admired a horse named Izzy and wanted to buy a colt that had Izzy as a parent.   Izzy's owner wanted to sell her Lucinda, a cute colt with big white spots, weak hooves and flaring nostrils.   Izzy has little white spots, strong hooves and normal nostrils.   Could Izzy be a parent of Lucinda?

2-11.   Mutants of beans were found that produced color variations in the seedlings. Strain 12 had pale green leaves, and strain 15 had lighter green leaves.   When these two strains were crossed in various ways, these were the results:

	Leaf Color			
Crosses	**Normal green**	**Lighter green**	**Pale green**	**Yellow-green**
12 X 12			all	
15 X 15		all		
(F$_1$) of a 12 X 15 cross	59			
F$_1$ X F$_1$ (F$_2$)	311	107	128	32
F$_1$ X 15	14	13		
F$_1$ X 12	20		17	

Define some gene symbols and then show the genotypes of strains 12 and 15 as well as the F$_1$ and the F$_2$.

*2-12. The locus *pax6* is a control gene that plays a role in embryological development of the eyes and face. Wishing to find more alleles of *pax6*, investigators treated a mouse strain with a mutation-causing substance and then inbred the mice. They watched for mice with defects in the face, and isolated nine mutant lines. Then crosses were done. Crosses could not use homozygous mutants since they were very severely affected and died rather early. (We will use a less cumbersome symbolism for the mutations, simply calling them *P2-P10* and calling the non-mutated alleles *p2-p10*.)

Wild type strain CS X Mutant strain 4 produced 64 wild type and 61 with the Mutant 4 phenotype. Some of the 61 mice from the first cross were then crossed to each other to produce five wild type mice, six moderately affected mutants and four severely affected mice.

Explain these results: How many loci are involved? Which allele is dominant? Why do you think there are two classes of mutant offspring in the second cross?

CHAPTER 3

PROBABILITY AND STATISTICS

[**Note:**   The material in this section on probability may be introduced at the end of Chapter 1 as the seventh step, "Step G."]

Some simple mathematical techniques are helpful in working genetics problems.

The Meaning of Probability

A **probability statement** tells how sure we are that something will occur. We have a day-to-day feeling for probability when we think about, for example, how sure we are that everyone in a class will receive an "A" grade or that the sun will come up in the east tomorrow. But to be useful to geneticists and other scientists, probability needs a quantitative meaning.

The **probability** of an event is the number of times t that an event will occur in n trials, or $p = t/n$. If the event's occurrence is absolutely certain, so that if we look n times we see the event n times, the probability is $p = n/n$ or 1 or 100%. If we are certain that the event will never occur, then $p = 0/n$ or 0. Probabilities are expressed as fractions, decimals or percents.

> **Solved Problem 3-1.** When the weather person says that there is a 20% chance of rain tomorrow, what does that mean?
>
> **Answer.** When weather conditions are like today's, 2 times in every 10 it rains.
>
> **Solved Problem 3-2.**   A geneticist collects 5000 flies and finds 25 with a gene for white eye color. If he goes out and collects again, what is the probability of finding a fly with white eye color if the conditions remain the same?
>
> **Answer.** The probability is $p = 25/5000 = 0.005$ or 0.5%

The concept of probability helps think about genetics ratios. When we say that the cross *Aa* X *Aa* produces 1/4 *AA,* 1/2 *Aa,* and 1/4 *aa,* we are making probability statements. We mean that in such a cross there is a probability of 25% that any offspring will be *AA* or *aa* and a probability of 50% that the offspring will be *Aa.*

> **Solved Problem 3-3.** A woman learned that she was heterozygous for the sickle cell allele, which results in sickle cell anemia when the allele is homozygous. It can be detected in the heterozygous condition by doing a test on the blood cells, and she and her husband both were found to be heterozygotes. The woman asked

what the probability was of their having a child with sickle cell anemia or a child carrying a copy of the allele. What should the doctor tell her?

Answer. Since both parents are heterozygotes, they have 1 chance in 4 (25%) of having a child with sickle cell anemia and 1 chance in 2 (50%) of having a child who carries a copy of the allele.

Using Probability in Genetics—the Multiplication Rule.

Once you grasp a few simple rules of probability, many problems more complex than those we solved in Chapters 1 and 2 can be done with ease, replacing the more cumbersome Punnett square method.

An **independent event** is one with a probability of occurring that is not influenced by other events being considered. This does not mean that nothing influences the event, but none of the other events being considered has any influence.

In the case of **mutually exclusive events,** if one event happens, the other cannot happen. For example, you can either get up at 8:00 a.m. or you cannot. If you do either one, the other will not occur. You can't both get up and not get up.

Solved Problem 3-4. Here is a list of one of two mutually exclusive events. Indicate what the other mutually exclusive event is.

A) Toss a coin and have it come up tails.
B) Only eat vegetables.
C) Be born on August 6.
D) Come in first in the Boston Marathon.
E) Witness an auto accident at the corner of Fifth and Washington Street.

Answer.
A) Toss a coin and have it come up heads.
B) Eat other foods besides vegetables.
C) Be born on any other day but August 6.
D) Come in any other time but first.
D) See no accidents or witness an accident at any place but Fifth and Washington or witness an accident involving mules at Fifth and Washington.

Questions about the probability of one independent event *and* another independent event are answered by multiplying the two probabilities. This is called the **Multiplication Rule.**

This is an important idea for genetics because it permits figuring out the probabilities of combinations of genes if the probability of each allele's appearance is known.

Independent assortment at two loci is a case of the simultaneous occurrence of alleles of those two loci appearing in the same gamete at meiosis. Since loci showing independent assortment are on different chromosomes, the segregation of alleles at one locus is not influenced by the segregation of alleles at the other locus. That is why we get a 9:3:3:1 ratio in a dihybrid cross. Consider an organism with the genotype *Aa Bb*, two unlinked loci, each heterozygous. Here is what happens if we consider the dihybrid cross as two monohybrid crosses:

Cross:		*Aa* X *Aa*			*Bb* X *Bb*	
Offspring		*A_* and *aa*			*B_* and *bb*	
Ratios		3/4	1/4		3/4	1/4

The probability of an *A_* is 3/4 as is the probability of *B_*. Being *B_* has no influence on what will happen at the *A* locus so the probability of the possible outcomes of both crosses is found my multiplying the probabilities together.

*A_ B_* is	3/4 X 3/4	or	9/16
A_ bb is	3/4 X 1/4	or	3/16
*aa B_* is	1/4 X 3/4	or	3/16
aa bb is	1/4 X 1/4	or	1/16

Solved Problem 3-5. In a cross of two *Aa Bb* individuals, what proportion of the progeny are *Aa Bb*?

Answer: You could do this problem with a Punnett square, but it is much more direct to do these problems with the Multiplication Rule. Consider each locus separately.

Cross:	*Aa* X *Aa*			*Bb* X *Bb*		
Offspring	*AA*	Aa	*aa*	*BB*	*Bb*	*bb*
Ratios:	1/4	1/2	1/4	1/4	1/2	1/4

So then the probability of *Aa Bb* is the product of the proportions of *Aa* and *Bb*.
1/2 X 1/2 = 1/4

PRACTICE PROBLEMS

3-6. Suppose there was a cross in which each parent was heterozygous for 10 different loci. What proportion of the offspring will be homozygous for all 10 recessive alleles?

3-7. A farmer said to a geneticist, "If I cross these 15 heterozygous black-and-white horned cows to a red-and-white bull that I'm pretty sure is heterozygous for the hornless gene, and I get two calves from each cow, how many of the calves can I expect to be red and white and hornless like their father?"

3-8. Consider Problem 1-35. Suppose two organisms of the genotype in that problem

(*Aa BB cc Dd Ee ff gg HH*) were crossed to one another. What proportion of the offspring would have the same genotype as the parents?

*3-9. This is a more challenging problem.
The fuzziness of cotton seeds is controlled by alleles at several loci.
- Allele *n2* is a recessive allele for lack of fuzz.
- The *n2n2* homozygotes have fuzzless seeds when a recessive allele (*n3*) at another locus is homozygous.
- *N2N2* and *N2n2* have fuzzy seeds.
- *N3N3* and *N3n3* do not permit *n2n2* to produce fuzzless seeds.

If two double heterozygotes are crossed, what proportion will be fuzzless? Use the **Product Rule of Probability** to find out. (Turley and Kloth, 2002)

When you combine probabilities:
- Frequency of **this *and* that** calls for **multiplication of probabilities.**
- Frequency of **this *or* that** calls for **addition of probabilities.**

Solved Problem 3-10. In a colony of mice kept for genetic experiments, a viral infection accounts for 7% of the deaths in a colony, and failures of the staff (forgetting to water the mice or dropping a cage, etc.) accounts for 1% of the deaths. Which kind of problem is it? Solve the problem.

Answer: The question asks what failure is due to this *or* that, viruses or humans. Therefore, this is a question about adding probabilities. 7% of the deaths plus 1% of the deaths is 8% of the deaths due either to viruses or human error.

Solved Problem 3-11. Look again at Problem 3-3. The chances of two heterozygous people having a child with sickle cell anemia is 1/4 or 0.25. The chance of a child not having the disease was 3/4 or 0.75.

A) What kind of problems are these and how do we solve them?
B) If they have two children, what is the probability that both of them will be homozygous for sickle cell anemia?
C) What is the probability that one of them will be a heterozygote and the other will be homozygous recessive?

Answer. A) These are problems using the Product Rule of Probability.
B) This involves independent events. The probability of one child having the disease is 0.25. Each child is an independent event. The phenotype or genotype of one child does not influence the phenotype or genotype of other children. So the question asks what is the probability of two children being born with sickle cell anemia? It is 0.25 X 0.25 = 0.0625 or 6.25% (or 1/4 X 1/4 = 1/16).

C) The probability of being a heterozygote is 0.5. The probability of being homozygous for sickle cell anemia is 0.25. The problem asks what is the probability of having one child of each of these kinds—one homozygote and one heterozygote? This and that. $0.5 \times 0.25 = 0.125$ or 12.5 %.

PRACTICE PROBLEMS

3-12. A man has sickle cell anemia (due to homozygosity of recessive allele s) and is normal for cystic fibrosis but is a carrier for it (because he is heterozygous for the c allele). A woman is normal but heterozygous for both loci. What is the probability of their having a girl child with both diseases? What is the probability of having a boy with a genotype just like the mother's?

3-13. The numerals 0 to 9 appear in the phone book in approximately equal amounts in the last position and the next-to-last place. What would be the probability of the last two numerals of a number chosen at random be either a 23 or a 44?

3-14. Chickens may have white or colored feathers. Feathers have color when a dominant allele C is present,. The cc homozygote results in white, Another locus is the dominant allele I, the inhibitor, which interacts with C. When the dominant I is present (II or Ii), production of color by C is blocked. Thus, only the genotype $ii\ C\_$ produces colored chickens. I and c are independently assorting. A third independently assorting locus produces black color when BB; WW if splashed white; and blue if BW.

Two white blue chickens heterozygous for all three loci were crossed. What proportion of their offspring will be white? Blue? Black?

Using Probability in Genetics: The Binomial Expansion.

Questions about the probability of one **group** of mutually exclusive events *or* another such **group** is answered by **expanding a binomial**. A binomial is an expression in the form $(p + q)^n = 1$. In our use of this expression p and q are the probabilities of mutually exclusive genetic events, like being tall or short, red-eyed or white-eyed.

This is an important idea for genetics because it permits figuring out the probabilities of various groups of phenotypes or genotypes if you know the probabilities of each individual genotype or phenotype.

Here is an example of the kind of problem for which expanding a binomial is useful. Suppose a couple wants to have children, but both are heterozygous for Tay-Sachs disease, a condition that appears in infants who are homozygous for the recessive allele. It is a disease with which an infant who has it suffers greatly. They want to know what their chances are of having four children without Tay-Sachs and the probability of having one child with the disease and three children without.

Notice the form of this problem. Probabilities of various combinations of mutually exclusive events are determined, such as "all four without the disease" or "three without and one with." In this case the important information is the probability of the event to occur or not to occur—in other words, independent events.

When you expand a binomial, you are multiplying $(p + q)$ by itself as many times as the size of the exponent. Thus to expand $(p + q)^3 = 1$ is $(p + q) \times (p + q) \times (p + q) = 1$, and $(p + q)^7 = 1$ is $(p + q) \times (p + q) \times (p + q) \times (p + q) \times (p + q) \times (p + q) \times (p + q) = 1$. Trying to multiply all the ps and qs is a confusing process prone to mistakes.

Here's what you do broken into small steps.

1. Recognize a problem as being one that can be solved with the help of the binomial expansion: problems that ask about probabilities of groups of individuals having one of two possible mutually exclusive events happen to them.
2. Note how many individuals will be considered. In the previous example, it was four children.
3. Now note that a binomial is in the form $(p + q)^n = 1$, where p and q are the probabilities of the two mutually exclusive events and n is the number of people in the group. For the example above, n = 4.
4. The number of terms that come from multiplying the **binomial** is one more than the number of individuals in a category. In our example, there would be $4 + 1 = 5$ terms.
5. For $(p + q)^n$, the first term will be p^n and the last one will be q^n. For the inside terms, there will be both p and q with **exponents** and a **coefficient**. An exponent is a number put in a mathematical expression behind and slightly above a letter to tell its power. For example, an expression might contain q^4. The "4" is an

exponent that tells us to raise q to the fourth power. A coefficient is a number put in front of a letter with which it is to be multiplied.

6. For the Tay-Sachs example, $(p + q)^4 = 1$ will be expanded. Start by writing the probabilities in this way:

   Lay out the terms in a line p and q on the outside.

7. Then add the inside terms: $p + pq + pq + pq + q$. I knew to put three pq in the middle because the total number must equal n +1 or 5 in this case.

$$\text{coefficient} \longrightarrow 4P^3 \longleftarrow \text{exponent}$$

8. Now put in the exponents, starting with the ones on p.

$$p^4 + p^3q + p^2q + pq + q$$

Then the exponents on q. $p^4 + p^3q + p^2q^2 + pq^3 + q^4$

9. Put the coefficients in the front of each term.

(For the end terms the coefficient is always 1.)

   For the inside coefficients, we begin the one next to the left end term.

   (step 1) In each case locate the exponent of the previous term

   (step 2) Multiply it by the exponent on the term being constructed.

   (steps 3 and 4) divide by the rank of the previous term. The rank of a term is its order in an expanded binomial (step 3 &4).

   (step 5) The coefficient is 4.

The Coefficient Cycle:
Follow the numbered arrows.

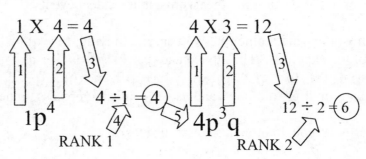

The next coefficient is in Rank 3. By the same procedure for 1, and 2, the term is (6 X 2/3) = 4. For Rank 5, the coefficient is 1 since it is the last term in the expansion.

The expanded binomial looks like this:

$$p^4 + 4p^3q + 6p^2q^2 + 4pq^3 + q^4$$

Solved Problem 3-15. Rather than just believe that the last term has an exponent of 1, use the same procedure to calculate the exponent of the fifth term as was used in the three inner terms.

Answer. The previous coefficient at Rank 4 is 4, and the exponent is 1. Remember: (Previous coefficient X Previous exponent/previous rank) = (4 X 1/4) = 1.

Solved Problem 3-16. Expand the binomial for n = 7; i.e., $(p + q)^7 = 1$.

Answer. Since n = 7, there will be n + 1 terms, which is 8. Lay out the *p*s and *q*s:

$$p + pq + pq + pq + pq + pq + pq + q = 1$$

Add the exponents:

$$p^7 + p^6q + p^5q^2 + p^4q^3 + p^3q^4 + p^2q^5 + pq^6 + q = 1$$

Add the coefficients:

$$p^7 + 7p^6q + 21p^5q^2 + 35p^4q^3 + 35p^3q^4 + 21p^2q^5 + 7pq^6 + q = 1$$

Each term of an expanded binomial is the probability of a particular combination of mutually exclusive events. This is the reward we get for figuring out all those coefficients.

For example, consider the earlier example woman who wanted to know the probably of having four children, only one of which had Tay-Sachs disease.

Four children are at issue, n = 4. The expanded binomial looks like this:
$$(p4 + 4p^3q + 6p^2q^2 + 4pq^3 + q^4) = 1$$

Let p = probability of a child having Tay-Sachs disease and q = probability of one child not having the disease. Since each parent is heterozygous and the gene for the syndrome is recessive, the probability of a homozygous child is 1/4, so p = 1/4. p + q = 1, so q = 1 – p = 3/4.

Now we can use the binomial we expanded. The first term gives the probability of four children all having Tay-Sachs, the second term gives the probability of three with Tay-Sachs and one without, the third term is for two with and two without, the fourth term is for one with and three without, and the final term is all without. The probability of three without the disease and one with it is, therefore, $4pq^3$ or 4 X 1/4 X $(3/4)^3$ then 4 X ¼ and $(3/4)^3 = 27/64 = 0.421875$, about 42%.

PRACTICE PROBLEMS

3-17. A molecular biologist has a group of centrifuge tubes to balance before putting them in the ultracentrifuge. Eight tubes all weigh the same and are heavy, while 12 other tubes all weigh the same and are lighter than the other eight. If she chooses two tubes at random to put in the ultracentrifuge, what is the probability that she will get one heavy and one light tube? Two heavy tubes, two light tubes?

3-18. For each term from an expanded binomial listed below, figure out what the term that follows it will be.

$3\ p^2q$ is the second term
$28\ p^2q^6$ is the seventh term
$25\ p^4q^3$ is the fourth term
$120\ p^7q^3$ is the fourth term

3-19. If the sex-ratio of humans is 1:1, among all families of six children, what number of boys will be most frequent?

3-20. A group of geneticists meets at a conference center of a university. Each table where people will sit to eat breakfast will seat four people. Assume that the sex ratio of this group is 60% men and 40% women. What proportion of the tables will have both men and women sitting there?

Statistics.

Statistics are numerical methods that may be used two ways in the analysis of data. **Descriptive statistics** make summary statements about data such as averages and variability.

Statistical inference allows us to make objective decisions when comparing groups of data. Statistical inference or "statistical tests" act like outside arbitrators between you and your data to help eliminate wishful thinking. Does your data fit your hypothesis? Statistical inference includes a set of rules set up ahead of time so the decisions will not depend on how much we like the hypothesis. This method uses the concept of probability to set a certain level of significance for an event, a statement of how probable it is that your hypothesis is confirmed by your data.

The Chi Square Test—Part 1

The probability associated with the statistic used in this section, the **chi-square** test is the **probability that the difference between the observed and expected ratios could have happened by chance alone**. This can be a little confusing, so take each part separately.

Suppose your hypothesis is that the single locus you are studying has two alleles, one dominant and one recessive. If your hypothesis is correct, a cross of two supposed heterozygotes should produce a 3:1 ratio. You then do the cross and get a ratio of 289:125. Is that close enough to a 3:1 ratio to confirm the hypothesis? How sure can you be that this is close enough? When you follow the directions below, you will get a probability, but a probability of what? .

1. You do the cross and get 289:125, your **observed** ratio. (Usually abbreviated "O".)

2. Your predicted ratio is found by a simple calculation.
   A) Determine the sum of all the organisms you obtained. That's 414.
   B) If your hypothesis is correct, 1/4 or 103.5 organisms would have the recessive phenotype, and 3/4 or 310.5 would show the dominant trait.
   C) So now you know the **expected** ratio, 310.5 : 103.5. (Usually abbreviated "E".)

   Solved Problem 3-21. If you thought a female insect you were studying was a heterozygote for a single dominant gene, and you did a test cross of that insect, what would your expected ratio be if you got 105 progeny?

   Answer. A test cross is a cross of an unknown genotype to a homozygous recessive, resulting in a 1:1 ratio of dominant to recessive if the unknown is a heterozygote. So half the 105 would be dominant and half would be recessive. The expected ratio is thus 52.5 : 52.5.

   Solved Problem 3-22. Suppose you had tomato seeds from a cross of two pure breeding tomatoes. You let the plants that grew from those seeds self-fertilize and you got four classes of these offspring totaling 250 individuals. The ratio is 135:47:53:15. What is your hypothesis, and what would be the complete expected ratio?

Answer. To find what proportions of the total organisms are in the least common category, divide the total by the number of individuals in this class. 250/15 = 15.6. This is very close to 16. In a 9:3:3:1 ratio, 1/16 are of the least common type homozygous for two loci, each with a dominant and a recessive allele. The complete expected ratio is 149.4 : 49.8 : 49.8 : 16.6.

The Chi-Square Test—Part 2

In the example just before Problem 3-21, the difference between the observed ratio and the expected one is - 21.5 or 21.5 fewer dominants than expected, and + 21.5 or 21.5 more than expected of the recessives. The difference you have obtained could be due to two different reasons:
1) Your hypothesis is incorrect.
2) Your hypothesis is correct, but there is some random variability introduced due to sampling error.

The greater the difference between observed and expected, the more probable that your hypothesis is incorrect. But how probable is that, and how do we know? This is where the chi-square (X^2) test comes in.

Step 1. Decide if chi-square is the appropriate test. This test is for determining "goodness of fit" between an observed and expected distribution. There are other kinds of statistical inference, but this is the one most often used for this purpose. Be sure that these requirements of the chi-square test are met:

Requirements for a X^2 (Chi Square) Test—Part 3
- Use only actual numbers, not frequencies or percent.
- The total number of individuals in the test should be greater than 30.
- No class should be smaller than five.
- No class may be zero.

Step 2. Calculate chi-square. Do that by the following steps:

A) Determine the observed ratio, O. This is simply the ratio you got in the cross. In the example before Problem 3-21, the observed ratio is 289:125. *You always use the actual numbers from the cross, not the frequencies.*
B) Determine the expected ratio. This depends on your hypothesis. In the example before Problem 3-21, the expected ratio is 310.5 : 103.5.
C) Use these numbers to calculate chi-square.

$$X^2 = \sum \frac{(O-E)^2}{E}$$ which in words means "subtract the

expected numbers in each category from the observed number (O − E), square the result $[(O-E)^2]$, divide the resulting number by the expected number in each category

$[(O - E)^2/E]$ and then add the results in each category together (that's what the Σ [pronounced "sigma"] in the equation is for)."

It is helpful to keep organized during the calculation, so use a form and fill in step by step from left to right. In math symbolism, this equation can be translated into a simple table to get the correct chi-square every time.

❶ Record your observed values for each category of progeny. In this case, there are two categories, dominant and recessive.

❷ Calculate how many you would expect in each class if your hypothesis is right.

❸ Subtract each expected number from its observed number.

❹ Square the result of step ❸.

❺ Divide each result in step ❹ by the expected number of that category.

❻ Add up all the numbers in step ❺. (That's what "Σ" means.)

❶	❷	❸	❹	❺
OBSERVED	EXPECTED	$(O-E)$	$(O-E)^2$	$(O-E)^2/E$
289	310.5	- 21.5	462.25	1.49
125	103.5	21.5	462.25	4.48
			❻ TOTAL (X^2)	0.938

Solved Problem 3-23. Look again at Solved Problem 3-21. Suppose the observed ratio was 50:55. Calculate X^2 for the cross involving the female mutant insect.

Answer: The expected ratio has already been calculated. Here is the completed table:

OBSERVED	EXPECTED	$(O-E)$	$(O-E)^2$	$(O-E)^2/E$
50	52.5	-2.5	6.25	0.119
55	52.5	2.5	6.25	0.119
			TOTAL (X^2)	0.238

Solved Problem 3-24. Calculate the X^2 for the results of the tomato cross described in Solved Problem 3-18.

Answer: The observed and expected ratios are given in 3-22. Here is the filled in table. Notice that two more rows have to be added for this cross.

OBSERVED	EXPECTED	$(O-E)$	$(O-E)^2$	$(O-E)^2/E$

135	140.6	- 5.6	31.36	0.223
47	46.9	0.1	0.001	0.000
53	46.9	6.1	37.21	0.793
15	15.6	0.6	0.36	0.024
			TOTAL (X^2)	1.04

Step 3. Find the probability in a chi-square table. A chi-square table is on pg. ???.
Using a chi-square table to find a probability value is a lot like using a computer: You don't have to know how a CPU works in order to use a word processing application to produce this sentence. Similarly, you don't have to know how a chi-square table works in order to decide which chi-square value corresponds to which probability.

Here is a diagram of the parts that make up a chi-square table.

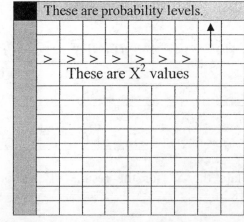

To look up a probability

1. Choose the appropriate degrees of freedom. This number is one less than the number of categories that you have. For a 3:1 ratio, there are two categories, so one degree of freedom. For a 9:3:3:1 ratio there are four categories, so three degrees of freedom, and for a ratio of 12:3:1 degrees of freedom is 2.

2. Starting at the left of the row associated with the correct degrees of freedom, move across the table until you come to the chi-square nearest the one you calculated. Then move up that column to find the probability associated with this chi-square.

Solved Problem 3-25. Refer again to Problems 3-21 and 3-23. We have an expected ratio of 52.5 : 52.5 and a chi-square of 0.238. What is the probability that the deviation from expected is due to chance alone?

Answer. There are two categories in this case, so the degrees of freedom is 1. As you move across the table from left to right, the chi-square values are 0.004, 0.016, 0.15, 0.46 and so on up to 10.83. A value of 0.238 is between 0.15 and 0.46. Now move up to the probability row and notice that you are between p values of 0.7-0.5.

Step 4. Decide what the result means. You have a probability, but what's the probability the probability of?

Look at the formula for X^2.
$$X^2 = \sum \frac{(O-E)^2}{E}$$

Two outcomes of a cross determine X^2: The bigger the $O - E$ is, the bigger X^2. So the bigger the chi-square, the less likely your hypothesis. The bigger the $(O - E)$ is, the larger will be the X^2. A difference of 20 between O and E will be more significant in a cross with 100 offspring than in a cross with 1000 offspring, so dividing by E takes sample size into account.

The probabilities on the X^2 table are the probability that the difference between the observed and the expected values is due to chance alone. "Deviation by chance alone" means the data suffer from random sampling errors—meiosis isn't perfect, and neither are geneticists, so variability creeps into the data. The more the observed deviates from the expected, the less likely it could have been a fluke of sampling.

Now how do you decide if the probability of deviation by chance alone is low enough to scuttle your hypothesis? There are many considerations, but most geneticists use a value for p of either 0.1 or 0.05,, accepting their hypothesis if the p value falls at or below their chosen cut-off. $p = 0.05$ means that one time in 20 you might get this ratio by chance.

Solved Problem 3-26. Two crosses were performed and chi-square tests were done. One cross produced a 60:40 ratio (1:1 was expected), and another produced 31 : 21 : 19:29. (1:1:1:1 was expected.)

Calculate the chi-squares for each cross. The number of the chi-square values are similar. What are the associated probabilities of each X^2? Why are the probabilities so different?

Observed	Expected	(O – E)	$(O-E)^2$	$(O-E)^2/E$
60	50	10	100	2

40	50	-10	100	2
			Total (X^2)	4.00

P associated with this chi-square is very nearly 0.05. One degree of freedom.

O	E	$(O-E)$	$(O-E)^2$	$(O-E)^2/E$
21	25	-4	16	0.64
31	25	6	36	1.44
19	25	- 6	36	1.44
39	25	-4	16	0.64
			Total (X^2)	4.16

Associated probability is between 0.20 and 0.03.

The reason for the different values of p from nearly equal chi-squares is that the degrees of freedom are different. In the chi-square table, the chi-square value decreases as the degrees of freedom increases.

Solved Problem 3-27. Two different crosses had different numbers of progeny so that one had a ratio of 840:860 and the other had a ratio of 40:60. A 1:1 ratio was expected. Calculate the X^2 for each cross. Even though they both deviated by 10 from the expected ratio, one had a much larger chi-square value than the other. Explain that result.

O	E	$(O-E)$	$(O-E)^2$	$(O-E)^2/E$
840	850	-10	100	0.118
869	850	10	100	0.118
			Total (X^2)	0.236

O	E	$(O-E)$	$(O-E)^2$	$(O-E)^2/E$
40	50	-10	100	2
60	50	10	100	2
			Total (X^2)	4

Answer. In calculating the chi-square, you divide by the sample size. So although both crosses had an O − E of 10 in each category, O − E was divided by 50 in one case and it was divided by 850 in the other. Dividing by the expected takes sample size into account since smaller samples have larger deviations.

Solved Problem 3-28. The tomato cross in 3-22 and 3-24 had a chi-square of 1.04. Give the associated p value and tell whether the data confirm or refute the hypothesis.

Answer. Use three degrees of freedom this time, since there are four categories. The probability associated with this chi-square of 1.04 is between 0.7 and 0.9. That is a long way from p = 0.05. The deviations can easily be ascribed to sampling error.

PRACTICE PROBLEMS

3-29. Here are the outcomes of various crosses involving *Paramecium tetraurelia*, a ciliated protist found widely distributed in fresh water. In each case, the data indicates what their hypothesis told them they should get from these crosses and also tells what they actually got. Calculate chi-square for each cross and tell whether to accept or reject the hypothesis. Each ratio involved a different locus and alleles.

A) Expect 1:1   Observe 115:129
B) Expect 1:2:1   Observe 69:115:50
C) Expect 1:2:1   Observe 69:110:83
D) Expect 1:2:1   Observe 42:80:29

3-30. Alleles for root color in radishes show no dominance: *WW* (white), *RW* (red), *RR* (purple). An independently assorting locus has two alleles for root shape, also with no dominance: *LL* (long), *LS* (oval), *SS* (round). Cross two *LS RW* radishes. Since there is no dominance, the progeny come in nine phenotypes. What is the ratio of F_2 radishes? Does the following distribution differ from the expected by more than chance alone? The observed ratio is already in the chart.

Ratio of F_2 radishes is:
1 Long Purple   *LL RR*
2 Long Red   *LL RW*
1 Long White   *LL WW*
2 Oval Purple   *LS RR*
4 Oval Red   *LS RW*
2 Oval White   *LS WW*
1 Round Purple *SS RR*
2 Round Red   *SS RW*
1 Round White *SS WW*

To help keep things straight, the ratio is inserted in column E.

O	E	(O – E)	$(O – E)^2$	$(O – E)^2/2$
20	1			
65	2			
40	1			
53	2			
100	4			
71	2			
37	1			
60	2			

34	1				
				Total	

3-31. Gregor Mendel calculated ratios arising from his crosses of garden peas, but he didn't know about the chi-square test since it hadn't been invented yet. Figure out if his data were consistent with his hypotheses for each of these crosses.

Mendel's hypothesis was that each of these first two was a monohybrid cross. The ratios of the two phenotypes in the F_2 are shown.

A) Tall X short

F_1 all tall | F_2 787 tall:277 short | Mendel's ratio 2.84 : 1

B) Green seeds X yellow seeds

F_1 all yellow | F_2 6032 yellow:2001 green | Mendel's ratio 3.01 : 1

C) Yellow round seeds X green wrinkled seeds | All F_1 round and wrinkled

Cross two F_1. and get 315 yellow round:101 yellow wrinkled:108 green round:32 green wrinkled. | What would be the hypothesis and is it supported by these data?

3-32. You have learned about several different ways to use probability in genetics. However, you won't benefit from them if you don't know which method to use in any given situation. Tell what method is useful in each problem stated below.

A) What percentage of our crop was damaged either by hail or trespassing picnickers?

B) Does this 9:6:1 ratio affirm the hypothesis that two genes and gene interaction are involved?

C) If a person has a five goats that are either spotted or non-spotted taken from a herd in which 40% of the goats are spotted and 60% are not, what is the probability that three of his goats are spotted and two are not? That two are spotted and three are not?

D) In a cross of *Aa Bb CC dd Ee* X *AA BB Cc Dd EE,* what proportion of the offspring will be like the parent on the right?

CHAPTER 4

MORE COMPLEX SITUATIONS

Many genetics problems go beyond the simplest situations due to complex relations between genotype and phenotype.

No Dominance

When there is dominance, the heterozygote has the same phenotype as the dominant homozygote. However, dominance is not always observed. Many genes show **incomplete dominance,** in which the heterozygote has a phenotype intermediate to the two homozygotes. For example, a heterozygote with alleles for red flowers and for white flowers might be pink. Other genes show **codominance**, another case of no dominance, in which the heterozygote has the phenotype of both homozygotes. For example, a person heterozygous for the A and B blood types is type AB, having both the A and B blood cell antigens. At first these two situations may seem the same, but they are distinct. Incomplete dominance is "in between" and codominance is the expression of both alleles. **Incomplete dominance and codominance make your task a little easier because all of the genotypes have unambiguous phenotypes.**

The symbols used when there is no dominance are sometimes slightly different from those used for cases of dominance. Often when there is no dominance, each allele will be given a number or different letter as a superscript. For example: R^1, R^2, etc., or I^A, I^B, etc. Lower case letters tend to be reserved for recessive alleles.

> **Solved Problem 4-1.** If you cross true-breeding four o'clock plants with red flowers to true-breeding four o'clock plants with white flowers, the resulting heterozygotes have pink flowers. What will be the ratio of red, white and pink flowers if two pink-flowered plants are crossed?
>
> **Answer.** Let R^1R^1 be red and R^2R^2 be white. Pink will be R^1R^2.
>
> F$_1$: Crossed $\qquad$ R^1R^2 $\quad$ X $\quad$ R^1R^2
>
> Gametes: $\qquad\qquad$ half the gametes from each parent will be R^1
> $\qquad\qquad\qquad\quad$ half the gametes from each parent will be R^2

	R^1	R^2	
R^1	R^1R^1 Red	R^1R^2 Pink	Ratio of 1:2:1
R^2	R^1R^2 Pink	R^2R^2 White	

The difference from a cross in which red would be dominant is that the pink flowers can be distinguished from the red flowers.

Solved Problem 4-2. A true-breeding radish with long red roots was crossed to a true-breeding radish with round white roots. The F_1 radishes were all oval and purple. Assuming that two independently assorting loci are involved, what would be the ratio of all the possible phenotypes among the F_2 produced by crossing two of the F_1 radishes? (See Problem 3-30 for more on this radish cross.)

Answer. Diagram the cross first.

P: Long Red   X   Round White

F_1:   All Oval Purple

F_2:   ??

We know that the parental radishes are all homozygous (true-breeding) for two independently assorting loci—probably one for color and one for shape. The F_1 is probably heterozygous for both loci. We know that the F_1 is different from either parental type, so it is probably a case of incomplete dominance for both loci. Let's choose some gene symbols:

"Shape" locus S^L = long S^N = round
"Color" locus C^R = red C^W = white

(Because no alleles are dominant, we don't use the upper and lower case symbols.) Now diagram the cross again with all that you know.

P: Long Red ($S^L S^L C^R C^R$)   X   Round White ($S^N S^N C^W C^W$)

F_1:   All Oval Purple ($S^L S^N C^R C^W$)

F_2:   ??

Now we are ready to answer the question. Cross two F_1 plants, determine the kinds of gametes they will produce and construct a Punnett square.

$S^L S^N C^R C^W$ will produce four kinds of gametes: $S^L C^R$, $S^L C^W$, $S^N C^R$, $S^N C^W$. The gametes are then arranged on a Punnett square to determine the genotypes of their offspring.

	$S^L C^R$	$S^L C^W$	$S^N C^R$	$S^N C^W$
$S^L C^R$	$S^L S^L C^R C^R$ Long Red	$S^L S^L C^R C^W$ Long Purple	$S^L S^N C^R C^R$ Oval Red	$S^L S^N C^R C^W$ Oval Purple
$S^L C^W$	$S^L S^L C^R C^W$ Long Purple	$S^L S^L C^W C^W$ Long White	$S^L S^N C^R C^W$ Oval Purple	$S^L S^N C^W C^W$ Oval White
$S^N C^R$	$S^L S^N C^R C^R$ Oval Red	$S^L S^N C^R C^W$ Oval Purple	$S^N S^N C^R C^R$ Round Red	$S^N S^N C^R C^W$ Round Purple
$S^N C^W$	$S^L S^N C^R C^W$ Oval Purple	$S^L S^N C^W C^W$ Oval White	$S^N S^N C^R C^W$ Round Purple	$S^N S^N C^W C^W$ Round White

The phenotypic ratios can be found by counting up the number of each phenotype in the Punnett square:

Long Red	1/16	Oval Red	2/16	Round Red	1/16
Long Purple	2/16	Oval Purple	4/16	Round Purple	2/16
Long White	1/16	Oval White	2/16	Round White	1/16

PRACTICE PROBLEMS

4-3. Decide whether each of the following descriptions is an example of dominance, incomplete dominance or codominance.

A) People with sickle cell anemia are homozygous for the *s* allele (*ss*). The homozygote *SS* does not suffer from sickle cell anemia. *Ss* heterozygotes are sometimes said to have "sickle cell trait." Under severe stress they may suffer some of the symptoms of sickle cell anemia, but to a much reduced degree of severity.

B) Two plants of grain amaranth with black seeds were crossed to each other. Among the progeny were three plants with black seeds for each plant with yellow seeds.

C) The D^I allele in horses dilutes the resulting color of some other coat color genes. The homozygote $D^I D^I$ dilutes more than the heterozygote $D^I D$.

D) Blood group M is due to the *M* allele in homozygous condition. Blood group N is due to the *N* allele in homozygous condition, and blood group MN is due to the *MN* heterozygote.

E) Tropical fish called "sword tails" are olive green in their natural habitat. A variant isolated by hobbyists is bright orange with black spots and is called "Montezuma." A single locus is involved, but a true breeding strain of Montezuma has not been produced because homozygous Montezuma is lethal—it results in dead fish. Heterozygotes are "Montezuma" and homozygous wild types are olive green.

4-4. Would the results of Problem 4-2 have been any different if we had started with true-breeding radishes that were long white and round red?

4-5. In cattle, the coat can be either red ($C^R C^R$), white ($C^W C^W$) or roan ($C^R C^W$). Roans have both white and red hairs. Horns are either normal (hh) or missing (HH or Hh). The coat color alleles show codominance, and the horn alleles show dominance. Suppose we crossed two $C^R C^W Hh$ cattle. What proportion of their offspring would have horns and be roan? What proportion will lack horns and be white?

*4-6. The *pax6* locus in mice affects the development of eyes and other parts of the face. Several new alleles of the locus were produced by treating mice with a mutation-causing chemical and then inbreeding. Homozygotes showed a much greater effect than did heterozygotes, but the investigators could classify genotypes, including heterozygotes, on the basis of their phenotype. Here are the results of crosses of one of these *pax6* alleles.

Parents:	Pp	X	pp		Pp	X	Pp	
	normal	moderate mutant			normal	moderate mutant	extreme mutant	
Progeny:	Pp	pp			PP	Pp	pp	
	83	67			13	32	12	

Is this a case of dominance, codominance or incomplete dominance? Give your reasoning. (Favor *et al.*, 2001)

Multiple Alleles

So far we have limited ourselves to loci with only two alleles, winged vs. wingless, striped vs. solid, disease vs. no disease. But there can be more than two alternate forms of one locus. A single locus represented by three or more alleles in a population is said to have **multiple alleles**. Solving genetics problems with multiple alleles is not that much more difficult than solving simpler problems since diploid organisms can still only have two alleles at any locus. The important thing is to be systematic. As long as you consider what is happening in only one individual at a time, every multiple allele problem becomes a collection of problems involving two alleles.

Solved Problem 4-7. It was first thought that the albino locus in rabbits had only two alleles, the wild type C and the recessive c. Recessive homozygotes lack melanin and have white hair and pink eyes. Several other alleles of this locus are now known. A slightly more complicated symbol system is used that adds a superscript to alleles. In addition to the dominant C and recessive c allele designations, the "chinchilla" (c^{ch}) and "Himalayan" (c^h) are used to unambiguously designate the multiple alleles for this locus. Any single rabbit inherits only two of these alleles, which segregate in a normal fashion during meiosis. Write the genotypes of the progeny that will result from each of the following crosses:

A) Cc X $c^h c^h$ B) Cc X $c^h c^{ch}$ C) cc X $c^h c^{ch}$

Answer. A) P: Cc X $c^h c^h$

Gametes: 1/2 C all c^h
          1/2 c

	C	c
c^h	Cc^h	cc^h

B) P: Cc X $c^h c^{ch}$

Gametes: 1/2 C 1/2 c^h
          1/2 c 1/2 c^{ch}

	C	c
c^h	Cc^h	cc^h
c^{ch}	Cc^{ch}	cc^{ch}

C) P: cc X $c^h c^c$

Gametes: all c 1/2 c^h
                     1/2 c^{ch}

	C
c^h	cc^h
c^{ch}	cc^{ch}

In Problem 4-7 we only looked at the genotypes. When considering phenotypes in multiple allele problems, dominance is a little more complicated. "Dominant" and "recessive" are relative terms that refer to what happens in a heterozygote. If there is dominance, the heterozygote will look just like the dominant homozygote. When there are multiple alleles, an allele may be dominant to one allele and recessive to another or show incomplete or codominance with one allele and complete dominance with another. Multiple allele problems are much easier if you first gather as much genotype information as you can and then work on the phenotypes.

Solved Problem 4-8. The human ABO blood group locus has three alleles, I^A, I^B and i. I^A and I^B are both dominant to i, but they show codominance in $I^A I^B$ heterozygotes. The ABO alleles control antigens on the surface of red blood cells that can be recognized with the appropriate antibodies. Persons whose cells have A antigen are said to be Type A, those with B antigen are Type B, those with both the A and B antigens are Type AB, and those without either antigen are Type O. Here is a chart of the relation of genotypes to phenotypes.

Blood Groups	**Genotypes**
A	$I^A I^A$ or $I^A i$
B	$I^B I^B$ or $I^B i$
AB	$I^A I^B$
O	ii

What are the expected phenotypes produced among the offspring of two people who are Type AB? Among the offspring of a Type A heterozygote and a Type B heterozygote?

Answer. First lay out what is known about the genotypes, work that out for parents and offspring and then deal with phenotypes.

Type AB people have the unambiguous genotype $I^A I^B$. The cross looks like this and is like any other cross we've considered with alleles lacking dominance.

P: $I^A I^B$ X $I^A I^B$

Gametes: I^A and I^B I^A and I^B

Punnett square (Fill in the genotypes and *then* worry about the phenotypes.):

First do genotypes   Then add phenotypes

	I^A	I^B
I^A	$I^A I^A$	$I^A I^B$
I^B	$I^A I^B$	$I^B I^B$

	I^A	I^B
I^A	$I^A I^A$ Type A	$I^A I^B$ Type AB
I^B	$I^A I^B$ Type AB	$I^B I^B$ Type B

In the second cross, all three alleles are involved, but only two are ever in any one individual.

P: $I^A i$ X $I^B i$

Gametes: 1/2 I^A 1/2 I^B
            1/2 I 1/2 I

Punnett square (Fill in the genotypes and *then* worry about the phenotypes.):

First do genotypes   Then add phenotypes

	I^A	i
I^B	$I^A I^B$	$I^B i$
I	$I^A i$	ii

	I^A	i
I^B	$I^A I^B$ Type AB	$I^B i$ Type B
i	$I^A i$ Type A	ii Type O

Solved Problem 4-9. Mrs. Idengaku was one of two mothers in a maternity ward. When she was given baby #1, she denied that it was hers, claiming baby #2 instead. The other mother also claimed baby #2.

Mrs. Idengaku is blood type O. Baby #1 is A, and Baby #2 is O. Unfortunately, Mr. Idengaku died just before the baby was born, so we can't find out his phenotype or genotype, but the Idengakus had three other children whose phenotypes are known. Keiko is Group A, Tohru is Group B and Kenichi is Group B. Assuming that Mr. Idengaku is the father of all four Idengaku children, who is right, Mrs. Idengaku or the other mother? What is your reasoning?

Answer. Set up the cross between Mr. and Mrs. Idengaku and put in all the information we have from the problem.

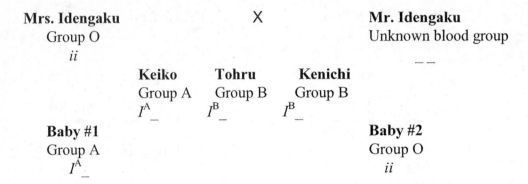

Mrs. Idengaku   X   **Mr. Idengaku**
Group O   Unknown blood group
ii $\_\,\_$

         Keiko   **Tohru**   **Kenichi**
         Group A   Group B   Group B
         $I^A\_$ $I^B\_$ $I^B\_$

Baby #1   **Baby #2**
Group A   Group O
$I^A\_$ ii

There are Idengaku children with both I^A and I^B alleles. Since Mrs. Idengaku is ii, those alleles must both have come from the father. So Mr. Idengaku was $I^A I^B$. If he was $I^A I^B$, the Idengakus cannot have had a child who is ii, since Mr. Idengaku has no i allele to give the baby. They could, however, have a child that is Group A. Their Group A children would be $I^A i$. Thus, Mrs. Idengaku is probably mistaken. Baby #2 could have been an Idengaku. Baby #1 could be an Idengaku, however.

PRACTICE PROBLEMS

4-10. With regard to ABO blood groups, one parent in a family is Group A and the other is Group B. If all four blood groups appear among the offspring, what proportion of the offspring will be Group O?

4-11. The rabbit coat color genes described in Problem 4-7 show a complex form of stepwise dominance. When the c (albino) allele is homozygous, it prevents pigment formation, so the homozygous individual has white hair. Allele c^h produces Himalayan rabbits with light bodies and dark noses. It is dominant to albino. The c^{ch} (chinchilla) allele produces a coat color that is lighter than wild type but darker than Himalayan, and it produces a light gray coat color when heterozygous with c or c^h. Allele C is dominant to all three of the other alleles. We can represent that as a dominant series $C > c^{ch} > c^h > c$. What are all the types of coat color that will be seen in the parents and offspring of a cross between a Cc^{ch} heterozygote and a $c^h c$ heterozygote?

4-12. A multiple allele series in domestic cattle influences the spotting pattern of the coat. Allele S produces "Dutch belt," s^h the Hereford pattern, s^c a solid color and s the Holstein pattern. These genes show an order of dominance in which S is dominant to the other three, s^h is dominant to all but S, s^c is dominant only to s, and s is recessive to all the other alleles (i.e., $S > s^h > s^c > s$).

A) Can Holsteins produce Hereford offspring?

B) Can Herefords produce Holstein offspring?

C) Can both solid color and Holstein cattle be offspring of one pair of Dutch belt cattle?

*4-13. In a study of the effect of certain alleles on the male-female sex ratio, investigators proposed that at one locus there were three alleles that differed from each other in the amount of testosterone each produced. L^H results in high levels, L^N has no effect on levels, and L^L produced lower than normal levels of testosterone.

A) Given these alleles, how many genotypic classes are possible?

B) L^H is dominant to both of the other alleles, and L^L is dominant to L^N. How many phenotypic classes are possible in those circumstances? (Astolfi, Cuccia, and Martinetti, 2001)

Gene Interactions: Epistasis

Sometimes alleles at one locus will have an effect on the expression of alleles at another locus. When a gene at one locus masks the expression of genes at another locus, the process is called **epistasis**. The result of epistasis is a modification of the phenotypic ratios you would expect if there were no epistasis. The genes still assort independently, but some of the phenotypes may be reduced in number or missing. You do these problems exactly as you would any other problem involving two or more genes. When you come to determining the phenotypes, however, you have to be more careful since you have to take those interactions into account.

Solved Problem 4-14. Black mice and brown mice differ in how the pigment melanin is distributed in the hair. Allele *B* is dominant and results in black hair, while allele *b* is recessive and results in brown hair when homozygous. Another locus determines whether the pigment melanin is synthesized at all. The recessive *c* allele blocks melanin production so the resulting "albino" mice have white hair. The dominant *C* permits melanin production. Obviously, if there is no melanin production due to *c*, it does not matter whether *B* or *b* is present. So allele *c* is epistatic to the *B/b* alleles when *c* is homozygous. What would be the phenotypes and genotypes of parents and their offspring if we crossed two mice heterozygous for both loci?

Answer. Diagram what we know already from the problem.

$$Bb\ Cc \quad X \quad Bb\ Cc$$

These mice have a *C* allele, so they can make pigment; and they have a *B* allele, so they will be black. Thus, each parent can produce four kinds of gametes (*BC, Bc, bC* and *bc)*, so a Punnett square will look like this:

	BC	Bc	bC	bc
BC	BB CC Black	BB Cc Black	Bb CC Black	Bb Cc Black
Bc	BB Cc Black	BB cc White	Bb Cc Black	Bb cc White
bC	Bb CC Black	Bb Cc Black	bb CC Brown	bb Cc Brown
bc	Bb Cc Black	Bb cc White	bb Cc Brown	bb cc White

If this cross of two heterozygotes had involved two loci, each with dominance but with no epistasis, we would expect a 9:3:3:1 ratio of phenotypes among the progeny. But epistasis by the homozygous albino allele masks the color gene. So:

*B_ C_*	Black	9/16	*bb C_*	brown	3/16
B_ cc	white	3/16	*bb cc*	white	1/16

That is 9 black:3 brown:4 white. We see the same gene arrangement as always, but different phenotypic ratios. The easiest way to do these problems is first to work out the segregation and independent assortment of genes and then to determine the phenotypes.

Solved Problem 4-15.   A gray horse was crossed to a chestnut horse. All the offspring were gray. Then two of the offspring (F_1) were crossed. Among the F_2, 12/16 were gray, 3/16 were bay, and 1/16 were chestnut. Explain these results by showing the genotypes of each phenotype among the parents, the F_1 , and the F_2.

Answer.   This problem is the reverse of Problem 4-14. This time you are being given phenotype data from which you have to figure out genotypes. In Problem 4-14 you were given the genotypes and how the genes interacted and were asked to figure out phenotypes. The secret of this second kind of problem is to find a ratio that you recognize and build on that.

Diagram what you know:

P: Gray   X   Chestnut

F_1:   All Gray

F_2:   12/16 Gray   3/16 Bay   1/16 Chestnut

Whenever you see sixteenths, you should think of the 9/16 : 3/16 : 3/16 : 1/16 ratio that results from crossing two organisms that are heterozygous for two independently assorting loci. In this case, the 9:3:3:1 ratio appears to have been modified by some sort of epistasis. One of the 3/16 classes has been combined with the 9/16 class.

Cross F_1:   *Aa Bb*   X   *Aa Bb*

F_2:   9/16 *A_ B_*   3/16 *A_ bb*   3 /16 *aa B_*   1/16 *aa bb*

The 1/16 category is unambiguously the double homozygous recessive. It always is, so chestnut has that genotype. It looks as if the *A* allele masks the expression of the *B* and *b* allele, so that whenever there is an *A* allele, the phenotype is gray:

9/16 *A_ B_* (Gray); 3/16 *A_ bb* (Gray); 3/16 *aa B_* (Bay); 1/16 *aa bb* (Chestnut)

And then, of course, the *aa B_* genotype produces the bay phenotype. In Problem 4-13, the epistatic allele (albino) was recessive. Here, the epistatic allele (gray) is dominant so we get a different modification of the 9:3:3:1 ratio. One parent was chestnut, so it must have been *aa bb*. To produce the double heterozygous F_1, the other parent must have been *AA BB*. We would expect that genotype to be gray, and it is!

A critical reader will wonder if the situation could be reversed. Could *B* mask the expression of the *A* and *a* alleles so that whenever there is a *B* the phenotype is gray? Then *aa B_* would be gray, and *A_ bb* would be bay. That is right. In this problem, it is purely arbitrary whether we call the epistatic gene *A* or *B*.

PRACTICE PROBLEMS

4-16. A 9:3:3:1 ratio occurs when two double heterozygotes are crossed. Gene interactions modify the 9:3:3:1 ratio. If a double heterozygote is test crossed (crossed to a double homozygous recessive), a 1:1:1:1 ratio is obtained when there is no gene interaction, but epistasis will alter this 1:1:1:1 ratio. For each of the following modified 9:3:3:1 ratios, what will be the modified 1:1:1:1 ratio if the double heterozygote is test crossed? For example, if the 9:3:3:1 ratio were modified to 9:3:4, the 4 class would be three single recessive homozygotes and the double recessive homozygote. So a test cross of the double heterozygote would yield 1:1:2.

A) 12:3:1   B) 9:7   C) 15:1

4-17. Because of the medical importance of mosquitoes, their genetics has been studied carefully, and many different genes have been found. In one species the larvae are ordinarily tan with a light-colored mid-dorsal stripe running along the abdomen. The allele *s* eliminates the stripe when homozygous. The *S* allele is dominant and results in the stripe. Another locus controls the body color, with the dominant allele *B* resulting in tan and the recessive *b* resulting in dark brown when homozygous. Allele *b* also reduces the mid-dorsal stripe to a tiny dot on the third segment of the abdomen. Suppose a homozygous normal mosquito is crossed to a dark brown mosquito with neither stripe nor spot. What will be the phenotypes of the F_1 and F_2 generations?

4-18. A black mouse was crossed to a hairless mouse. Since it lacked hair, it was impossible to tell if its hair was black or brown. All the progeny had black hair. When these F_1 progeny were mated among themselves, 100 F_2 were black, 43 were hairless, and 30 were brown. What were the genotypes of the original parents?

4-19. Geneticists who wanted to learn more about cucumber genetics began to search for genetic variation in the plants. They found genes for the color of the spines on the fruit.

The spines can be either black or white, and the inheritance pattern they observed was most simply explained as due to two independently assorting loci. At least one dominant allele had to be present at each locus for spines to be black. Otherwise spines were white. If they crossed two double heterozygotes and obtained 1000 offspring, how many plants would have white-spined fruit and how many would have black?

*4-20. Seeds of cotton are ordinarily covered with fuzz and fibers, but alleles have been found which result in seeds that are fuzzless (lack fuzz but have fibers) or fiberless (lacks both fuzz and fibers). A one-locus model had been proposed to explain the difference between strain 143 (fuzzless) and 5690 (has both fuzz and fibers), with fuzzless recessive to having fuzz and fibers, but the results of recent crosses favor a two-locus model. Here is a comparison of the two models.

One-locus model		Two-locus model	
strain 5690	strain 143	strain 5690	strain 143
N^2N^2	n^2n^2	N^2N^2 N^3N^3	n^2n^2 n^3n^3

According to the two-locus hypothesis, N^3N^3 or N^3n^3 blocks the effect of n^2. The n^3n^3 homozygote must be present for nn^2 to be expressed.

Show what ratios of genotype and phenotype would be expected for the F_2 of a cross of strain 5690 and strain 143 and the back cross of the F_1 to strain 143 under both the one-locus model and the two-locus model. (Turley and Kloth, 2002)

*4-21. The herbicide triallate has been used for years to control the weed *Avena fatua* (a wild oat). However, increased use of high levels of triallate has resulted in appearance of a triallate-resistant strain. A sensitive strain grown for a long time in the lab was crossed to the resistant strain. Ordinarily this plant self-fertilizes, so elaborate measures have to be taken for crossing. Therefore, they did not study resistance in the small number of F_1, only in the F_2. The results were:

F_2:   Sensitive   1088   Resistant   51   Total   1139

The investigators concluded that two loci were involved, each with two alleles. Show how that explains the data. (Kern *et al.,* 2002)

CHAPTER 5

SEX LINKAGE

Up to this point, all of the problems in this guide have involved autosomal genes, (i.e., genes that are not located on sex chromosomes). Genes on sex chromosomes add another dimension to genetics problems.

Some organisms have a **chromosomal method of sex determination** in which the two sexes have differences in one pair of chromosomes. In humans and fruit flies, females and males differ in a single pair of chromosomes called the **sex chromosomes.** In the females the two sex chromosomes are both **X-chromosomes**, while in the males there is an X-chromosome and a **Y-chromosome.** Often the Y-chromosome contains few genes while the X-chromosome contains many genes, so there are loci on the X that are not on the Y. The genes on the X that are not on the Y are called **X-linked genes.** (Sometimes they are called **"sex-linked" genes**, but, strictly speaking, sex linkage includes both X-linked genes and rare cases of Y-linked genes.) Problems involving sex linkage require a slightly different approach.

Use Informative Symbols

Problems involving X-linked genes are not difficult to do if you remember that the X- and Y-chromosomes segregate to gametes as any other chromosome pair would segregate and if you use a system of symbols to remind you of sex linkage. A simple way to do this is to write an X or a Y for the chromosome and then write the X-linked alleles as if they were connected to the X. For example:

A female fly heterozygous for the white eye allele (w): $X^W X^w$
A male fly with the white eye allele: $X^w Y$
A female human heterozygous for muscular dystrophy: $X^D X^d$
A male human with the allele for red-green color blindness: $X^c Y$

Notice that females can be either heterozygous or homozygous, but males have only one copy of the gene, so those terms can't be applied. Instead we say that males are **hemizygous.**

Solved Problem 5-1. Define symbols and write the genotypes of each of the following individuals.
A) A female fruit fly heterozygous for the X-linked allele for forked bristles.
B) A female human heterozygous for the recessive hemophilia allele.
C) A female human homozygous for the hemophilia allele.
D) A male human with the muscular dystrophy allele.

E) A male human without the allele for red-green color blindness.

Answer. First write the sex chromosome composition of each individual and then hook on the appropriate allele symbol.

A) A female, so XX. Let F be straight bristles and f be forked bristles. $X^F X^f$

B) A female, so XX. Let H be normal and h be hemophilia. $X^H X^h$

C) A female, so XX. $X^h X^h$

D) A male, so XY. Let the allele for muscular dystrophy be d. $X^d Y$

E) A male, so XY. Let c be the color blindness allele and C its normal allele. $X^c Y$

PRACTICE PROBLEMS

*5-2. Spermatogonia are the cells in testes that give rise to the sperm. When looking for genes expressed specifically in the spermatogonia, 25 loci were found, of which 10 on the X-chromosome and three others on the Y-chromosome were picked for further study. The Y-linked genes were not on the X-chromosome. How would you symbolize a Y-linked allele called b located on the Y-chromosome? A) In a diploid spermatogonium? B) In a haploid gamete? (Wang *et al.*, 2001)

Keep the alleles attached to their chromosomes.

> The X- and Y-chromosomes segregate into gametes much as a pair of alleles would. Half the sperm will receive a Y-chromosome and half will receive an X. All the eggs will receive an X. If the female is heterozygous at an X-linked locus, half of her eggs will contain an X with one of the alleles on it, and half will contain an X with the other allele on it. Remember that, and you can't go wrong.

Solved Problem 5-3. The first allele discovered by *Drosophila* geneticists was the recessive white eye allele *w*, which is X-linked. The normal eye is red. A red-eyed male was crossed to a white-eyed female, and a white-eyed male was crossed to a red-eyed female. All parents are from true-breeding lines. What kinds of progeny will be found in each case?

Answer. Define symbols and diagram each cross. Let X^W be the symbol for a chromosome bearing the red-eyed allele and X^w for one bearing the white-eyed allele. Remember that the Y has neither allele, so it is just Y.

	P:	Red male $X^W Y$	X	White female $X^w X^w$		White male $X^w Y$	X	Red female $X^W X^W$

Gametes:	1/2 X^W	all X^w		1/2 X^w	all X^W
	1/2 Y			1/2 Y	

	X^W	Y
X^w	$X^W X^w$ Red Female	$X^w Y$ White Male

	X^w	Y
X^W	$X^W X^w$ Red Female	$X^W Y$ Red Male

Solved Problem 5-4. In *Drosophila,* forked bristles (*f*) is an X-linked characteristic. Straight bristles (*F*) is dominant. Vestigial wings (*v*) is autosomal recessive, and its wild type allele (*V*) is dominant. A male with normal bristles and homozygous for normal wings is crossed to a vestigial winged female that is heterozygous at the forked locus. Show the genotypes and phenotypes of their offspring.

Answer. Outline the cross.

A)   P: *VV* $X^F Y$ X   *vv* $X^F X^f$

	Male	Female
	Normal Wings	Vestigial Wings
	Normal Bristles	Normal Bristles

Gametes: ½ $V\mathrm{X}^F$ ½ $v\mathrm{X}^F$

                  ½ $V\mathrm{Y}$ ½ $v\mathrm{X}^f$

	$V\,\mathrm{X}^F$	$V\mathrm{Y}$
$v\,\mathrm{X}^F$	$Vv\,\mathrm{X}^F\mathrm{X}^F$ Female Normal Wings Normal Bristles	$Vv\,\mathrm{X}^F\mathrm{Y}$ Male Normal Wings Normal Bristles
$v\mathrm{X}^f$	$Vv\,\mathrm{X}^F\mathrm{X}^f$ Female Normal Wings Normal Bristles	$Vv\,\mathrm{X}^f\mathrm{Y}$ Male Normal Wings Forked Bristles

All the females and half the males have normal wings and normal bristles. Half the males have normal wings and forked bristles.

PRACTICE PROBLEMS

5-5. Red-green color blindness is a recessive X-linked character in humans. In cases involving two parents with normal color vision, can male offspring be color blind? Can female offspring?

5-6. Look at the results of Problem 5-4. If one of the male offspring with forked bristles is crossed to a female heterozygous for both loci, how many of each possible type would you expect if there were 400 progeny from that cross?

5-7. Hemophilia is a defect in blood clotting due to a recessive sex-linked allele. Alexis, son of the Russian Tzar and Tzarina, Nicholas II and Alexandra, had hemophilia. Rupert Viscount Trematon of England, son of Alice of Athlone, also had hemophilia. The royal families of England and Russia were related. From this information, is it more probable that Alice was a cousin of Nicholas II or of Alexandra? Explain.

*5-8. Problem 5-2 described the Y-linked allele *b*. Could a person with allele *b* have a mate who also had *b*?

You can sometimes recognize the patterns of X-linked inheritance.

As with the autosomes, some problems tell you about the parents and ask you about the offspring; but some problems give you information about the offspring and expect you to recognize that the pattern of inheritance reveals sex-linkage. Several clues suggest sex-linkage. Not all of these clues will be found in every case, but any of them should cause you to consider sex-linkage as a possibility.

Clue 1. Among the progeny, the male phenotypes are in different ratios than the female progeny. Look, for example at the cross in Problem 5-4. The bristle gene is X-linked, and the vestigial gene is not. Both the male and female progeny have normal wings, but some of the male progeny have forked bristles while all of the female progeny have normal bristles. On the other hand, notice Problem 5-6. The males and females have the same phenotype for both wing and bristles. The presence of one of these clues is good evidence for sex-linkage, but the absence of clues does not eliminate the possibility of X-linkage. Sometimes you have to look at several different crosses to be sure.

Clue 2. Reciprocal crosses may give different results. In reciprocal crosses, the phenotype of the male in one cross is the phenotype of the female in the other. Problem 5-3 involves reciprocal crosses of the white eye gene. Notice that in one cross all the offspring have red eyes, but in the reciprocal cross the males have white eyes.

Solved Problem 5-9. Suppose a recessive allele l causes death in the early embryo so that homozygotes for l are never counted among the progeny. What would be the effect if the gene were autosomal and a female heterozygous for l were crossed to a normal homozygous male? What if the gene were X-linked and a heterozygous female were crossed to a normal male?

Answer.	If Autosomal			If X-linked	
P:	LL	X	Ll	$X^L Y$ X $X^L X^l$	
	Male		Female	Male Female	
Gametes:	all L		½ L	½ X^L ½ X^L	
			½ l	½ Y ½ X^l	
F_1	½ LL;		½ Ll	¼ $X^L X^L$; ¼ $X^L X^l$	
				Females all normal	
				¼ $X^L Y$; ¼ $X^l Y$ (dead)	
	All normal			Half the males missing	
	No difference in the two sexes				

So when this lethal gene is X-linked, there will be half as many male progeny as female progeny. There is no effect on the sex ratio when the gene is autosomal.

Solved Problem 5-10. Red-green color blindness is a defect in the color sensitivity of the retina. It was noticed that it tended to appear in males more frequently than in females and that a male with red-green color blindness could have two normal parents. Usually men with red-green color blindness do not have color blind children, but they can have color blind grandchildren. What kinds of children and grandchildren would you expect in a rare case of a woman with red-green color blindness, assuming that she married a man without the condition?

Answer. That two normal people can have a child with red-green color blindness is evidence that the gene is recessive. That a man with red-green color blindness could have affected grandchildren but no affected children suggests sex-linkage.

This is because the father is $X^c Y$. All of his daughters will get the X^c, but his sons will get the Y. If the mother is homozygous, none of their sons or daughters will be color-blind. But grandsons who receive the X^c through the daughter will be.

If red-green color blindness is recessive, a woman with the condition must be homozygous $X^c X^c$.

Among the children of a color blind woman, all the boys will get an X^c and will be color blind. All of the girls will get the normal X^C from the father and an X^c from the mother and will not be color blind. The female offspring are heterozygous and are called **carriers** for this recessive allele.

The sons of a color blind woman are $X^c Y$. If they have children by homozygous wives, none of these children will be color blind because all get an X^C from their mother. All of the daughters of a color blind woman, however, will be $X^C X^c$, If they have children by normal husbands, half of their sons would be expected to be color blind.

PRACTICE PROBLEMS

5-11. The Bar eye allele in *Drosophila* results in a reduction in the number of units in the compound eye so that the eye is long and narrow instead of round. This allele is dominant to the allele for normally shaped eyes. When males from a true-breeding Bar-eyed strain are crossed to normal females, the female offspring have Bar eyes, but the males do not. When normal males are crossed to females from a true-breeding Bar-eyed strain, all the progeny have Bar eyes. Explain this result.

5-12. As indicated in Problem 5-3, white eye in *Drosophila* is an X-linked allele. A geneticist wondered if another gene (for body color) was X-linked also. She crossed black-bodied males to white-eyed females. From the results below, can you decide if the body color gene is X-linked? Explain.

P: red-eyed, black-bodied male X white-eyed, yellow-bodied female

F₁: males: white-eyed, yellow bodied
           females: red-eyed,   yellow-bodied

F₂: males: females:
           red yellow   3/16   red yellow   3/16
           white yellow   3/16   white yellow   3/16
           red black   1/16   red black   1/16
           white black   1/16   white black   1/16

5-13. Here is a cross involving three different independently assorting loci. One of the loci is X-linked. Which one?

P: Long, Purple, Active, Male   X   Short, Yellow, Sluggish, Female

F₁: 1/4 Long, Yellow, Active, Males
           1/4 Short, Yellow, Active, Males
           1/4 Long, Purple, Active, Females
           1/4 Short, Purple, Active, Females

*5-14. Look again at Problems 5-2 and 5-8, involving a Y-linked allele. Can we tell if allele *b* is dominant or recessive? Can allele *b* be either homozygous or heterozygous?

CHAPTER 6

PEDIGREES

Pedigrees are another way to analyze inheritance (and they give you a chance to review the basic principles of solving genetics problems).

Humans present a special problem to geneticists because controlled crosses are not possible (or desirable). The numbers of offspring are often small, so working with ratios is not always possible either. As a systematic way of making the most of what data are available, geneticists use **pedigrees**, charts of several generations of inheritance. In pedigrees you examine past "crosses," looking for appearances of phenotypes like those of the individual under study and watching for a pattern.

Individuals are symbolized by circles and squares.

The pedigree focuses on the traits of individual people. Males are symbolized with a square, females with a circle, and people of unknown gender with a diamond. People who have the particular phenotype being studied have their circle or square filled in. Fetuses not carried to term or babies that were still-born (and thus are not available to determine the phenotype) are represented with a small dot.

Male	Female	Unknown Gender	Has the phenotype	Phenotype unknown
□	○	◇	●	•

Solved Problem 6-1. Suppose we were studying the condition called cystic fibrosis. Write the symbol for a man with cystic fibrosis and for a woman without it.

Answer. Man with cystic fibrosis: ■   Woman without cystic fibrosis: ○

Relationships between individuals are represented by lines.

A mating between two people is symbolized by drawing a horizontal line to connect them. Offspring of a mating are connected by a vertical line extending from the horizontal line between the parents, and siblings are recognized by a horizontal bracket. Usually progeny are represented from left to right in order of their birth. For example, here is a couple that produced a daughter first and then a son.

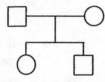

Solved Problem 6-2. A man marries and has two sons and a daughter. Then his wife dies and he marries again, having one daughter by the second marriage. Draw a pedigree that illustrates this scenario. (Hint: Connect the man to both of his wives.)

Answer. First wife 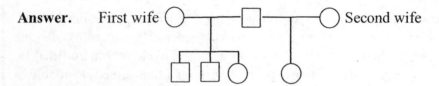 Second wife

PRACTICE PROBLEMS

6-3. Two people without Tay-Sachs disease give birth to two healthy girls and then to a boy with Tay-Sachs disease. Draw that pedigree.

6-4. Lionel and Leona met at a singles dance and married soon afterward. They were initially attracted to each other because they both had "free" earlobes—that is, earlobes not directly attached to their heads. After a respectable interval, they had a child who to their surprise had "attached" earlobes. Draw a pedigree to show Lionel, Leona and their child, whose name is Louella.

6-5. Two people with polydactyly produced a boy who also had the condition. That child eventually married and his wife (who does not have the condition) gave birth to a girl without the condition. Draw the pedigree that shows this scenario.

Pedigrees can represent the genetic history of many generations.

As Problem 6-6 will show, it is possible to include any number of generations on a pedigree and to show matings between any number of people on each generation. When more than two generations are studied, usually each generation is numbered with a Roman numeral, and each person in any one generation is numbered with an Arabic numeral. That makes it easier to refer to an individual.

Solved Problem 6-6. A pedigree follows several generations of a family afflicted with a hereditary disability. Those who have the disability are symbolized with a filled-in symbol. Answer the following questions using the reference

A) Which individuals have the disability?
B) Who are the grandchildren of III-4?
C) Which married couple are second cousins (i.e., the children of cousins)?

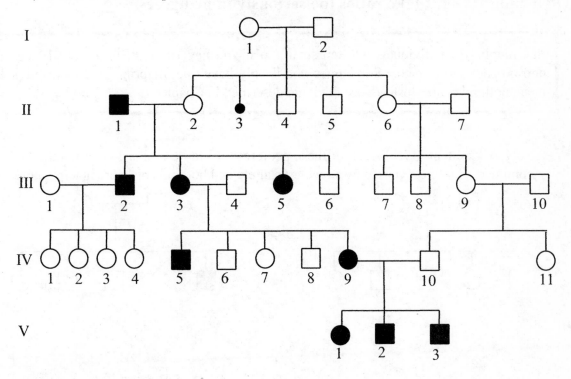

Answer.
A) All those with filled symbols have the disability: II-1, III-2, III-3, III-5, IV-5, IV-9, V-1, V-2, V-3.

B) Grandchildren are two generations after grandparents, so V-1, V-2 and V-3 are the grandchildren of III-4.

C) IV-9 and IV-10 are second cousins.

Analyze pedigrees using the same logic used for crosses.

Pedigrees represent crosses, although they are not controlled crosses. So we can use the same reasoning to learn from pedigrees that we use to learn from crosses. A pedigree is a description of phenotypes, so pedigrees can be difficult to analyze when you are given phenotypes and you need to determine information about genotypes. Usually pedigree problems ask, "What kind of inheritance pattern does this gene show?" (In other words, given the evidence in this pedigree, is the gene dominant or recessive, autosomal or sex-linked?)

Here are some principles for analyzing pedigrees.

Principle 1. Don't take ratios too seriously in pedigrees.

> The number of individuals is often small, so if you have four offspring, a 3:1 ratio doesn't mean too much. Sometimes you'll only have one offspring, and then ratios mean nothing at all. Instead, use the logic described in Chapter 1, Steps C to F.

Solved Problem 6-7. Which of these pedigrees could be produced if a family had a condition of interest caused by a recessive allele and both parents were heterozygous?

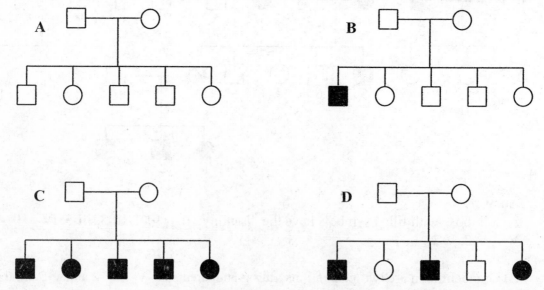

Answer. All four of them are possible if the parents are both heterozygous. If you depend on a ratio to answer this question, you might be led astray.

Principle 2: Is the condition of interest dominant or recessive?

Several rules of thumb will help you make this decision when looking at a pedigree:

If the condition is **recessive**, then

A) expression of the trait often skips generations (the recessive allele is transmitted via heterozygotes and is not expressed unless two heterozygotes produce a homozygous recessive child).

B) as long as the allele in question is fully expressed in every individual who has the allele, two affected parents always have affected children (two homozygous recessives can give rise only to homozygous recessive children); but unaffected parents may have affected children (two heterozygotes may have a homozygous recessive child).

If the condition is **dominant**, then

A) the condition shows up frequently, often in every generation (both homozygotes and heterozygotes express the phenotype).

B) affected parents may have unaffected children (two heterozygotes may have a homozygous recessive child); but unaffected parents never have affected children.

Solved Problem 6-8. The skin of individuals with xeroderma pigmentosum (a hereditary disease) is extremely sensitive to sunlight, often becoming thickened, freckled and cancerous. Based on the following hypothetical pedigree chart, can you say whether the disease is caused by a dominant or recessive allele?

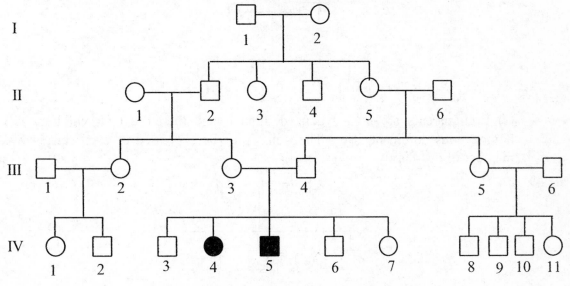

Answer. The condition must be recessive—it skips generations and appears among the offspring of unaffected parents, who must both be heterozygotes. (Note that these heterozygous parents are cousins—xeroderma pigmentosum is rare.)

Solved Problem 6-9. Albinism is an absence of pigmentation determined by an autosomal recessive allele. Which of the children of the following crosses would be affected by this trait?

A) Albino parents, five children: girl, girl, boy, girl, boy.

B) Albino mother and homozygous non-albino father, five children: boy, boy, girl, girl, girl.

C) Homozygous non-albino parents, five children: boy, girl, boy, boy, girl. Draw pedigrees to illustrate each of your answers.

Answer: A) Both parents are homozygous recessive, and therefore their children will be the same—homozygous recessive. Because albinism is a recessive condition, the whole family is affected, and the pedigree looks like this:

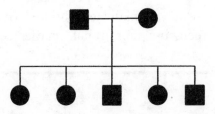

B) The parents are homozygous for different alleles, and therefore their children will be heterozygous—unaffected, but carriers of an allele for albinism.

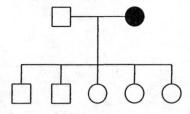

C) Both parents are homozygous dominant, and therefore their children will be homozygous dominant also. The trait won't be expressed or even carried by any member of this family.

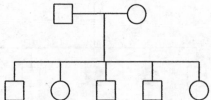

PRACTICE PROBLEMS

6-10. Look at the pedigree in Problem 6-4. Is the character of free earlobes dominant or recessive to attached earlobes?

6-11. Genetic pituitary dwarfism is an autosomal recessive condition. Here is a pedigree in which a man's deceased first wife (the crossed-out circle) was not affected by pituitary dwarfism, but his second wife is. Is the man heterozygous for this trait or homozygous dominant? Explain why.

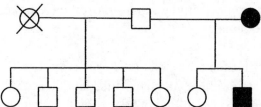

6-12. Look at the pedigree you drew in Problem 6-5. Can you tell if polydactyly is dominant or recessive? Explain. Suppose the couple in the second generation had another daughter who had polydactyly. What would that tell you about dominance?

6-13. Look at the pedigree in Problem 6-6. Is this condition due to a dominant or recessive allele? Explain why.

6-14. Again look at Problem 6-6. Which of the following individuals is probably heterozygous for the condition? For which is there not enough evidence to decide? II-1, III-3, III-6, IV-9, V-2?

*6-15. Dogs are susceptible to eye diseases called "progressive retinal atrophy," or PRA. In an attempt to characterize PRA in mastiffs, an affected female was crossed to a normal laboratory beagle without the PRA gene. From this pedigree, decide if the PRA allele is dominant or recessive. Why is the use of the lab beagle necessary for an unambiguous determination?

(Kijas *et al.,* 2003)

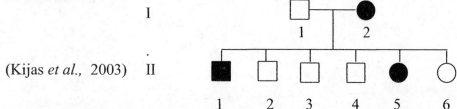

*6-16. This is the pedigree of a family, some of whose members have Noonan syndrome. I-2 has gray shading to indicate an unknown phenotype.

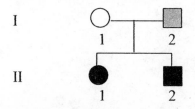

Can Noonan syndrome be due to a dominant allele? If so, what is the most likely phenotype for I-2?

Can Noonan syndrome be due to a recessive allele?   If so, what is the most likely phenotype for I-2?   (Tartaglia *et al.*, 2002)

Principle 3. Is the characteristic being studied autosomal or sex-linked?

There are some rules of thumb about how to tell if a gene is sex-linked.   Assuming that sex-linked genes are only on the X-chromosome (X-linked) and not on the Y and that males are XY and females XX,

If the gene is **dominant**, then
A) a father will pass the allele to all of his daughters and none of his sons, and everyone who receives a copy will have the associated phenotype.

B) a female who is heterozygous for a dominant X-linked allele will pass the allele to half of her offspring regardless of sex.

If the gene of interest is a **recessive,** then
A) males who have the gene will express it, since those males are hemizygous.   Thus recessive sex-linked phenotypes are more common among males than among females.

B) a man who has the gene will pass it on to all of his daughters and none of his sons. Since the gene is recessive, usually no offspring will express the gene, since most of these conditions are rare, making it unlikely that both parents would carry the gene. But in the *next* generation, the daughters will pass the gene to half of their sons. So a pattern in which the trait appears in the males of alternating generations may indicate X-linkage.

Solved Problem 6-17. Here is a pedigree in which the phenotype of interest appears in all the shaded individuals. Is the allele for the phenotype of interest dominant or recessive, and X-linked or autosomal? Explain.

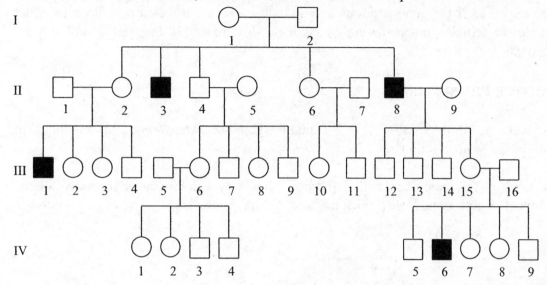

Answer: Children of unaffected parents are affected, so the allele must be recessive. It only shows up in males and an affected male (II-8) does not pass the gene to his offspring, so the allele is probably X-linked.

Solved Problem 6-18. Here is a pedigree for yet another condition. Is the allele for the shaded phenotype dominant or recessive, X-linked or autosomal?

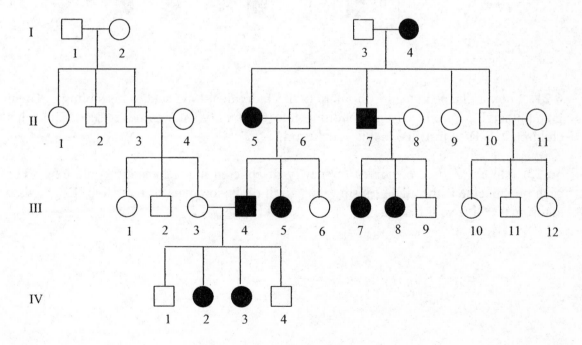

Answer. The phenotype only shows up when a parent has it too, so the responsible allele is probably dominant. More information would be available if there were one mating of two affected individuals to see if any unaffected offspring result—as would be expected if the gene is dominant. Females who are affected pass it on to either males or females, but males only pass it on to females. So the trait is probably X-linked.

PRACTICE PROBLEMS

6-19. Refer to Problem 6-17. Is I-1 homozygous or heterozygous at the locus in question? How do you know?

6-20. Look at this pedigree. Is the condition caused by a dominant or recessive allele? Sex-linked or autosomal? How do you know?

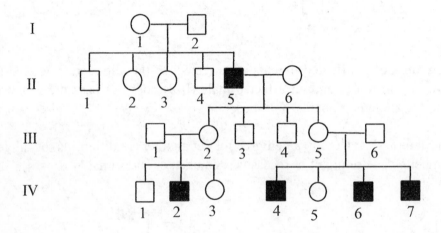

6-21. Look at the pedigree for Problem 6-20. Individual IV-6 said to his siblings, "Great grandfather I-2 introduced this condition into the family—it was never seen before his children." Is IV-6 justified in this conclusion?

*6-22. Here are three pedigrees for newly discovered human genetic conditions. For each one, decide if the allele causing the condition is dominant or recessive, X-linked or autosomal. Give your reasons.

A) **Bilateral Perisylvian Polymicrogyria.** Affects the structure of the cortex of the brain. Results in palsy, epilepsy and mild mental retardation. It is the most common of the polymicrogyrias, occurring at a frequency of one in every 2600 live births. Here are two pedigrees of two unrelated families. (Villard *et al.,* 2003)

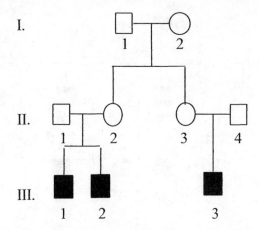

B) **Bilateral Frontoparietal Polymicrogyria.** Affects different parts of the brain from the condition in the previous pedigrees. The following pedigree involves some cousin marriages, symbolized with a double connecting line. (Piao *et al.,* 2003)

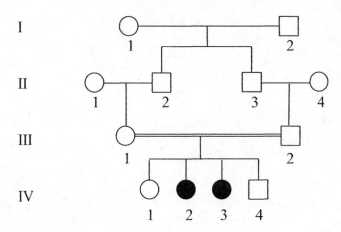

C) Susceptibility to pancreatic cancer. Pancreatic cancer is the fifth leading cause of cancer death. More than 29,000 cases are diagnosed each year. Some people, as in this pedigree, have a gene that increases the probability of having pancreatic cancer. (Eberle *et al.,* 2002)

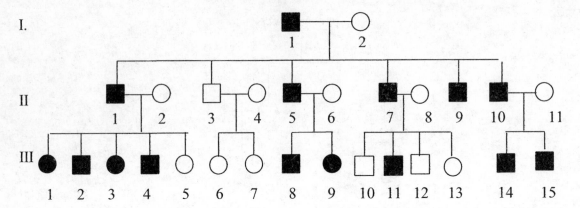

CHAPTER 7

LINKAGE

Some genes are linked together on the same chromosome.

So far all of the problems in this guide involve either single genes (or loci) or several independently assorting loci. However, organisms must have more genes than chromosomes. From the sequence information coming from the Human Genome Project, *Drosophila melanogaster* has four pairs of chromosomes when diploid but may have 12,000 to 16,000 genes. Humans have 23 pairs of chromosomes but 20,000 to 25,000 different genes. We know that the genes are located on the chromosomes, so sooner or later geneticists had to encounter genes on the same chromosome. When they figured out this aspect of genetics, they pictured each chromosome as a line of genes, each in a specific location in the line. Since they were thinking of these genes in terms of their locations, each gene in its location was called a **locus**. Two loci are said to be **linked** or to show **linkage** when they are on the same chromosome. Linkage brings new wrinkles to solving genetics problems, but be systematic and remember the basics and you will have no more difficulty than with any other genetics problems.

Use symbols that show the genes are linked.

Use a symbol system that shows that genes are on the same chromosome. If the alleles *A* and *B* are not linked, simply write the diploid genotype like this: *AA BB* or *Aa Bb*. But if they are linked, use a little line to symbolize the chromosome they have in common: *AB/AB* or *AB/ab*. Sometimes people use a double line to emphasize the diploid nature of the genotype, but that is not necessary. Because the genes are linked, the symbol *AB/ab* is different from *Ab/aB*. Both are double heterozygotes, but in the first case *AB* came from one parent and *ab* from the other. In the second case, *Ab* came from one parent and *aB* from the other.

Solved Problem 7-1. Symbolize the genotype for each of the following cases.

A) In *Drosophila*, flat wings *(C)* is dominant to curly wings *(c)*, and yellow ocelli *(W)* is dominant to white ocelli *(w)*. (Ocelli are little light-sensing organs on top of the head.) These two loci are on the same chromosome. Assume that in a heterozygote, one parent provided the chromosome with the allele for curled wings on it and the other parent provided the allele for white ocelli.

B) The same situation as in (A), but one parent provided alleles for both curled wings and white ocelli, and the other parent provided the two wild type alleles.

Answer. A) One parent provided *C* and *w*, while the other provided *c* and *W*. The symbol for this is *Cw/cW*.

B) When one parent provides both mutant alleles, the symbol is *CW/cw*.

Solved Problem 7-2. Sometimes crosses involving three linked genes are done. Yellow body (*y*), miniature wings (*m*), and forked bristles (*f*) are all linked on the X-chromosome. No alleles for these loci occur on the Y-chromosome. How would you write the symbol for these three genes in a female who is heterozygous for all three and received the mutant alleles yellow and forked from one parent and the mutant miniature from the other parent?

Answer. Using capital letters for the wild type alleles and lowercase letters for the mutant, this fly would be *yfM/YFm*.

PRACTICE PROBLEMS

7-3. Suppose you had a cat that was heterozygous for two loci on Chromosome 3, one for the presence (*N*) or absence (*n*) of the enzyme nucleoside phosphorylase and one for the presence (*H*) or absence (*h*) of the enzyme hexokinase A. If a female cat is *nH/Nh*, and her parents were both homozygotes, what were the genotypes of her parents?

7-4. Write the genotype symbols of one of the offspring of this cross:
  ABc/ABc   X   *abC/abC*

7-5. Mice have fuzzy hair when the recessive "fuzzy" allele (*f*) is homozygous, and they display a strange stumbling sideways motion when the recessive "tumbler" allele (*t*) is homozygous. These genes are linked on mouse Chromosome 8 A fuzzy mouse was crossed to a tumbler, each of which was from a pure line. Use appropriate symbols to show the genotypes of the animals in this cross.

*7-6. People with an "exercise intolerant allele" have a defect in an enzyme of lipid or carbohydrate metabolism, so that their muscles do not work well enough for exercise. Here is a table regarding these alleles.

Carbohydrate Metabolism		Lipid Metabolism	
Enzyme	**Chromosome**	**Enzyme**	**Chromosome**
Glycogen phosphorylase	11	Carnitine palmitoyl transferase	1
Phosphorylase kinase	X-chromosome	Acyl-CoA dehydrogenase	17
Phosphofructokinase	12		
Phosphoglycerate kinase	X-chromosome		
Phosphoglycerate mutase	7		
Enolase	17		
Lactate dehydrogenase (LDH)	11		

A) Which of these genes may be linked to one another? Why do you think so?
B) Define a gene symbol for each linked allele and write the genotype of each pair of linked loci for an individual with one parent homozygous for one of the mutant alleles and the other parent homozygous for the other mutant allele. (Rankinen *et al.*, 2002)

Detect linkage between two genes as a lack of independent assortment.

> Independent assortment happens when genes are on different chromosomes. Which allele of a particular locus goes into a gamete has no influence on which allele of a second or third independently assorting locus goes into that same gamete. But if the genes are connected to one another on the same chromosome, the occurrence of one allele in a gamete will have very much to do with the occurrence of a linked allele at another locus on that same chromosome.

For example, consider the cat in Problem 7-3, with a genotype of *nH/Nh*. At meiosis, *n* is on the same chromosome as *H*, so they should end up in the same gametes. *N* and *h* should occur together in different gametes. So in this case, the cat could produce two kinds of gametes, *nH* and *Nh*. In contrast, if the genes were assorting independently, four gamete types would occur in equal numbers: *nH, Nh, NH,* and *nh*.

> **Solved Problem 7-7.** Look at the two genotypes in Problem 7-1 (A) and (B). Show what kinds of gametes each would produce if the genes were completely linked together on a chromosome and what gametes they would produce if the genes were on different chromosomes and assorting independently.
>
> **Answer.** The genotype in (A) is *Cw/cW*, so the gametes produced would be *Cw* and *cW*. In (B) the genotype is *CW/cw*. This time the gametes are *CW* and *cw*. If the genes were assorting independently, each of these genotypes would produce these four gamete types in equal proportion: *CW, cW, Cw,* and *cw*.

> **Solved Problem 7-8.** Look at the genotype of the fly in Problem 7-2. Show what gametes it would produce (A) if the genes were completely linked and (B) if the three loci were independently assorting.
>
> **Answer.** The genotype *YFm/yfM* would produce two kinds of gametes if the loci were linked, *YFm* and *yfM*. If the genes were independently assorting, there would be $2^3 = 8$ different gametes, and they would be *YFM, YFm, YfM, Yfm, yFM, yFm, yfM,* and *yfm*.

Notice that in each case in which the genes are completely linked, the gametes that are produced are "**parental**" gametes. They have the same combinations of alleles as the parents did. **A clear excess of parental types among the offspring of a cross involving more than one allele is a sign of linkage.**

PRACTICE PROBLEMS

7-9. Three alleles are suspected of being linked. A parent of genotype *AA BB CC* is crossed to a parent of genotype *aa bb cc*. (We won't symbolize them as linked until we know for sure.) What kinds of gametes can each parent produce, and what is the genotype of the progeny of this cross? Consider both the case of independent assortment and complete linkage of the genes.

7-10. What gametes would the progeny from Problem 7-9 produce if all three loci are linked?

7-11. In *Drosophila* the recessive gene for vermilion eye is inherited as we would expect if it were X-linked. Likewise, the recessive gene for miniature wings is inherited as if it were X-linked. Are the two genes linked?

*7-12. Look at Problem 7-6. From the data on exercise intolerant alleles, we know that some of the loci are linked.

Glycogen phosphorylase may be normal (*G*) or exercise intolerant (*g*). That locus is linked to lactate dehydrogenase, which may be normal (*L*) or exercise intolerant (*l*).

Suppose there was a mating between a person homozygous for the exercise intolerant phosphorylase allele and a person homozygous for the exercise intolerant lactate dehydrogenase allele. What would their genotype be with respect to these loci? What would their offsprings' gametes be if these loci were tightly linked? (Raininen *et al.*, 2002)

Genes are usually not completely linked.

So far, in the problems you've solved involving linked genes, the two alleles stay together so that wherever one allele goes the other goes. This complete linkage happens in some cases, but it is more common to see some recombination. This happens because the two homologous chromosomes exchange material, a process called "**crossing over**." A physical basis for crossing over is probably the chiasmata seen at meiosis. If the genes are linked, there will still be an excess of the parental types, but some recombinant (**crossover or non-parental**) types will also be seen because of crossing over. The excess of parental types is a very good indication of linkage, even if there are recombinants as well.

Solved Problem 7-13. Return to Problem 7-1 to remind yourself about the genes for curled (vs. flat) wing and white (vs. yellow) ocelli. A cross between a homozygous fly with flat wings and yellow ocelli and a fly with curled wings and white ocelli resulted in heterozygous offspring. The male heterozygotes were crossed to females with curled wings and white ocelli to obtain another generation. Likewise, the female heterozygotes were crossed to males with curled wings and white ocelli. There is no crossing over in male *Drosophila* (which is not typical of most organisms).

A) Diagram the P and F_1 generations of the cross involving **male** heterozygotes, showing the phenotypes and genotypes of each fly involved, and then show the kinds of gametes the male F_1 will produce.

B) Diagram the P and F_1 generations of the cross involving **female** heterozygotes and show what kind of gametes the F_1 female produces. Unlike in the male, her chromosomes sometimes undergo crossing over. In this example, 20% of the gametes are recombinant or crossover types, and 80% are parentals.

Answer.

A) P: *CW/CW* male   X   *cw/cw* female
           flat wings   curled wings
           yellow ocelli   white ocelli

    F_1:   all *CW/cw*
                                flat wings
                                yellow ocelli

Male gametes (no crossing over, so no recombinants)

                        1/2 *CW*   1/2 *cw*

B) The diagram of the P and F₁ is the same as in part (A), but the gametes produced by the F₁ are different.

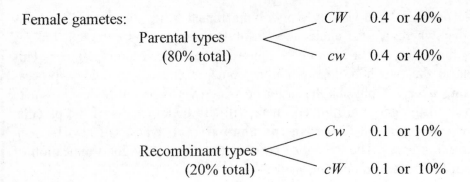

Female gametes:

Parental types
(80% total)

CW 0.4 or 40%

cw 0.4 or 40%

Recombinant types
(20% total)

Cw 0.1 or 10%

cW 0.1 or 10%

PRACTICE PROBLEMS

7-14. Suppose that both male and female F₁ are to be crossed to *cw/cw* (a test cross). The table below shows the genotypes that could result from the test cross. Fill in by entering the proportion of the offspring that will be of the genotype in the left-most column. Enter a "0" in all the cells in the table which would not be seen. Enter a frequency in each cell which would be seen.

You can get information about the proportion of crossovers from Problem 7-3. What will be the types of offspring and the frequency of each type of offspring from the cross using the male heterozygote? The female? What would be the types and frequency of offspring from either cross if there were independent assortment?

Genotypes of Test Cross Offspring	Male Heterozygote	Female Heterozygote	If Independent Assortment Happened
CcWw			
Cc ww			
ccWw			
Cc wv			

*7-15. In the plant *Lotus japonicus,* investigators have studied recombination between many genes spread throughout the genome. Loci *s5* and *s13* are linked on Chromosome 2. Heterozygotes for these two loci give recombination 30% of the time. If you had 500 plants, how many would there be of each type if the parents are *S5 S13/s5 s13*?

*7-16. Refer to Problem 7-15. If the parents were *S5 s13/s5 S13*, how many of each type would there be?

Note for Problems 7-15 and 7-16. The symbols for alleles in this problem are simplified from the paper where the data were found. (Pedrosa *et al.,* 2002)

A test cross makes it easier to detect recombination from independent assortment.

A test cross is the best way to detect deviation from independent assortment. Remember that in a test cross one parent is homozygous recessive for all the genes being considered. Because all the gametes from the test cross parent have only recessive alleles, the kinds and proportions of offspring from a test cross are the same as the kinds and proportions of gametes from the other parent. For example, if the other parent is *Aa Bb*, its gamete types will be *AB, Ab, aB* and *ab*, the *aa bb* parent will have all *ab* gametes and the offspring will be *Aa Bb, Aa bb, aa Bb* and *aa bb*.

Solved Problem 7-17. Do a test cross of each of the following heterozygotes and show the expected types and proportions of progeny if the genes assort independently.

A) *Aa Bb*   B) *RR Ss TT Ww Zz*

Answer. A)   *Aa Bb*   X   *aa bb*

Gametes: *A B   A b*   all *ab*

             a B   a b

Progeny: *Aa Bb*   The four types will occur in a ratio of 1:1:1:1.
         aa Bb
         Aa bb
         aa bb

B)   *RR Ss TT Ww Zz*   X   *rr ss tt ww zz*

Gametes: *RSTWZ   RsTWZ*   all *rstwz*
         RSTWz   RsTWz
         RSTwZ   RsTwZ
         RSTwz   RsTwz

Progeny: *Rr Ss Tt Ww Zz*   *Rr ss Tt Ww Zz*
         Rr Ss Tt Ww zz   *Rr ss Tt Ww zz*
         Rr Ss Tt ww Zz   *Rr ss Tt ww Zz*
         Rr Ss Tt ww zz   *Rr ss Tt ww zz*

A modification of the test cross approach can be done with X-linked genes. In that case, males are hemizygous. So if you cross a heterozygous female to a male with the recessive allele on its X-chromosome, the offspring will be in the same proportions as the gametes of the female. In effect, for X-linked genes, crossing a heterozygous female to a recessive male has the same usefulness as a test cross.

	Female		Male
	AB/ab	X	*ab*/Y

or you can write these: $X^{AB} X^{ab}$ $X^{ab} Y$

Gametes: X^{AB} (Parental) X^{ab}
X^{ab} (Parental) Y
X^{Ab} (Recombinant or crossover)
X^{aB} (Recombinant or crossover)

	X^{ab}	Y
X^{AB}	$X^{AB} X^{ab}$	$X^{AB} Y$
X^{ab}	$X^{ab} X^{ab}$	$X^{ab} Y$
X^{Ab}	$X^{Ab} X^{ab}$	$X^{Ab} Y$
X^{aB}	$X^{aB} X^{ab}$	$X^{aB} Y$

See how the proportions of the different genotypes and phenotypes are the same as the proportions of gametes from the heterozygous female?

Solved Problem 7-18. In *Drosophila* only 3% of the gametes show recombination between vermilion (a recessive eye color allele) and miniature (a recessive wing shape mutant). The two loci are both on the X-chromosome. F_1 females from a cross of a vermilion, miniature parent and a homozygous red-eyed, normal-winged parent, were crossed to vermilion, miniature males. What will be the proportions of each phenotype among the offspring? Consider males and females separately.

	Female heterozygote		Male for test cross
	VM/vm	X	*vm*/Y

Gametes: Parental types (97%)
    VM 0.485   *vm* 0.485   *vm*   Y
Recombinant (crossover) types (3%)
    Vm 0.015   *vM* 0.015

Female offspring: *VM/vm* 0.2425   Male offspring: *VM*/Y 0.2425
                 vm/vm 0.2425   *vm*/Y 0.2425

(The 97% parentals are divided among four classes: two classes of males and two classes of females.)

Vm/vm 0.0075   *Vm*/Y 0.0075
vM/vm 0.0075   *Vm*/Y 0.0075

(The 3% recombinants are divided among four classes also.)

PRACTICE PROBLEMS

7-19. How would the answer for Problem 7-18 differ if the original female heterozygotes were *Vm/vM*?

7-20. You can construct the female heterozygote for Problem 7-18 in two ways, starting either with *VM/VM* females and *vm*/Y males or with *vm/vm* females and *VM*/Y males. These females are identical, regardless of how they are constructed. Still, only one of the two possible crosses will also produce males that can be mated to the heterozygous females in a test cross. Which way works for doing the X-linked equivalent of a test cross?

*7-21. When people collect a large number of loci in an organism, they find that linked genes come in sets. For any one locus, many other loci will show independent assortment, but a group of other loci show the signs of linkage. For example, the plant *Antirrhinum majus* has eight linkage groups, and *Lotus japonicus* has six linkage groups. Given this information, what do you think the diploid number of chromosomes is in each of these species? Why? (Pedrosa *et al.,* 2002; Schwarz-Sommer *et al.,* 2003)

CHAPTER 8

LINKAGE MAPS

The frequency of recombination or crossing over between genes can be used to measure how far apart the loci are on their chromosome.

As you will see in the following examples, each pair of genes we look at will have its own unique frequency of crossover types. Imagine chromosomes as strings of beads with each gene locus in its characteristic location as one of the beads. Diploid organisms will have two of each of the chromosomes, one from the male parent and one from the female, with each bead at its place as a locus on the chromosome. At meiosis (See Chapter 9) a chromosome from the father and one with the same loci from the mother line up and trade lengths of their beads.

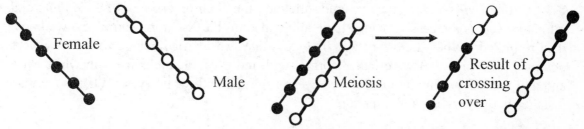

Now imagine that the exchange of chromosome beads can happen with equal likelihood all along the chromosome. If so, the farther the two beads (gene loci) are apart, the more crossing over will happen between them. Since the genes are arranged linearly on a chromosome, we can use the frequency of crossing over to make a "linkage map" of linked genes.

The unit of distance between linked genes is the percent (%) crossing over.

A road map shows the distance between two points in miles or kilometers. Treasure maps show distances in paces. Linkage maps use map units as their unit of distance. **One map unit** is defined as **1% crossing over**. You can calculate the % crossing over by doing a test cross of a heterozygote for two linked genes. The % crossing over is (the number of crossover types) ÷ (the total number of offspring) X 100.

Solved Problem 8-1. Many genes reside on the X-chromosome of *Drosophila*, including those for cut wings (*ct*), vermilion eyes (*v*) and miniature wings (*m*). In the appropriate cross, crossovers occur 3.1% of the time between *m* and *v* and 13% of the time between *ct* and *v*. Which wing gene is closest to gene *v*?

Answer. The distance between miniature and vermilion is much smaller (3.1 map units) than the distance between cut and vermilion (13 map units). So *m* is closer to *v* than *ct* is.

Solved Problem 8-2.   A recessive allele (*a*) in corn increases resistance to grasshoppers relative to the resistance of plants with the *A* allele, and another recessive allele at another locus (*z*) causes stripes on the leaves unlike the solid colored leaves of those with the *Z* allele.   A homozygous *aa ZZ* plant was crossed to a homozygous *AA zz* plant, and then the double heterozygote was test-crossed. Of 5280 corn plants resulting from this cross, 2480 were *aZ/az*, 2536 were *Az/az*, 140 were *AZ/az* and 124 were *az/az*. What is the map distance between these two genes?

Answer. Since the parents were *aZ/aZ* and *Az/Az*, *AZ/az* and *az/az* represent the crossover types. Of 5280 plants, 264 are recombinant. 264/5280 = 0.05, and 0.05 X 100% = 5%. The map distance is 5 map units.

PRACTICE PROBLEMS

8-3.   White (*w*) is a recessive eye color allele on the X-chromosome of *Drosophila*, and miniature (*m*) is a recessive wing shape allele also on the X-chromosome. So we know they're linked. When a cross of *wm/WM* X *wm*/Y was carried out, 1290 progeny were obtained, of which 843 were either normal with red eyes, or miniature with white eyes, and 447 were either miniature with red eyes or normal with white eyes. What is the map distance between white and miniature?

8-4. If the cross *wM/Wm* X *wm*/Y were done instead of the cross described in Problem 8-3, what proportion of the progeny would be miniature with white eyes or normal with red eyes? Why?

*8-5. A plant called *Lotus japonicum* is being studied for its agricultural potential. A map of the genes so far identified included five loci on Chromosome 1. Each locus is placed on the map according to its distance from the end of the chromosome. The distance from the end is measured in % recombination. So a gene placed at 39% would have 39% recombination with another gene located right at the end of the chromosome. These maps are made by totaling up the % crossing over between genes that are close together. Here are the distances of each of the loci from the end of the chromosome:

N1 32 m.u. C2 23 m.u. L 15 m.u. P 7.5 m.u. G2 6.5 m.u.

What % recombination would you expect in the heterozygote in each of the following crosses?

A) N1 C2/ n1 c2   X   n1 c2/n1 c2

B) L p/ l P   X   l p / l p

(Pedrosa *et al.,* 2002)

A map of a chromosome can be made by doing a set of pair-wise crosses.

If we have three linked genes, we can make a map that places all three in their relative order along the chromosome and that shows the map distances (% crossover) between the genes. An orderly strategy for building a gene map from data on crosses of pairs of genes will help you solve most problems.

1) **Determine the % crossing over between all three pairs of genes involved.**
2) **The pair with the longest map distance must be on the outside.**
3) **The two shorter distances are the distances from the outside genes to the middle one.**
4) **Check the order to see if the total distance between each outside gene and the middle gene is approximately equal to the distance between the outside genes.**

Solved Problem 8-6. In a certain organism, crossing over between R and S occurs 15% of the time, and crossing over between S and T occurs 22% of the time. Crossing over between R and T happens 7% of the time. What are the order and map distances for these three genes?

Answer.
Step 1. In this case, we know the genes are linked since they all yield a small percentage of crossover types.
Step 2. The largest map distance is 22 map units between S and T, so those are the outside genes.
Step 3. R must be in the middle, 15 units from S and 7 units from T. Here is the map:

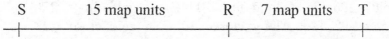

       S   15 map units   R   7 map units   T

Step 4. The sum of SR and RT is 22, which is what was observed, so this probably is the map.

Solved Problem 8-7. Pair-wise crosses involving three different *Drosophila* mutations on Chromosome 2 produce three heterozygotes, which are then test-crossed.

The cross involving black and vestigial resulted in 20,153 flies, of which 3578 were recombinant. The cross involving black and purple produced 48,931 progeny of which 3026 were recombinant. Finally, the cross involving vestigial and purple produced 2598 recombinants out of 13,601 flies. Map these genes.

Answer.

Step 1. The % crossing over in each case:

Black and vestigial 3578 ÷ 20153 X 1005 = 17.8%.

Black and purple 3026 ÷ 48931 X 100 = 6.2%.

Vestigial and purple 2598 ÷ 13601 X 100% = 19.1%

Step 2. Vestigial and purple are farthest apart at 19.1 map units, so they are the outside genes.

Step 3. The distances to the outside genes from the middle give this map:

Vestigial   17.8 map units   black   6.2 map units   purple

——————————+————————————————————+———————————————+——

Step 4. The sum of the short distances is 17.8 + 6.2 = 24, slightly more than 19.1, but it's close.

PRACTICE PROBLEMS

8-8. For these three pairs of linked genes, construct a linkage map showing the order and distance between the loci.

r and s   10% recombinants
s and t   20% recombinants
r and t   15% recombinants

8-9. In *Geneticus impossibili*, an organism found only in genetics problems, the antennae may be long or short; the eye facets may be gold or silver; and the most posterior bristles may be set close together or far apart. A homozygous long, gold, far-set individual was mated to a homozygous short, silver, close-set one. The resulting hybrids were test-crossed. The data were then divided up into pair-wise categories like this:

long, gold 296   long, far   444   gold, far   270
long, silver 198   long, close 50   gold, close 228
short, gold 202   short, far   50   silver, far   224
short, silver 304   short, close 456   silver, close 278

Construct a gene map.

*8-10. *Pax6* is a mutation which plays a role as a control gene for embryonic development of the eye. Newly arisen alleles of the *Pax6* locus all mapped to very nearly the same place on Chromosome 2. Two other Chromosome 2 loci, D2 and Agouti, were also used to determine where *Pax6* alleles were located. A heterozygote for all three of these mutants was test-crossed to produce this map:

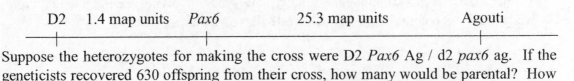

D2   1.4 map units   *Pax6*   25.3 map units   Agouti

Suppose the heterozygotes for making the cross were D2 *Pax6* Ag / d2 *pax6* ag. If the geneticists recovered 630 offspring from their cross, how many would be parental? How many would arise from a crossover between D2 and *Pax6*? How many would arise from a crossover between *Pax6* and Agouti? (Favor *et al.,* 2001)

An easier way to make a map is to do a test cross involving three linked genes.

Constructing a map of three genes from pair-wise crosses is not as economical as doing a single cross involving all three genes. Usually the cross is a test cross of the appropriate heterozygote for the linked genes and is called a three-point test cross. Below is a method for constructing a map from three-point test cross data.

1) **Are all three loci linked?**
2) **What is the map order?**
3) **What are the map distances?**

(As with other problems you've encountered in this book, you can analyze such crosses by breaking them down into simpler problems: three problems using pairs of linked genes. If the methods for analyzing all three genes at once confuse you, you can't go wrong if you treat these as three separate problems and then combine them.)

Are all three loci linked? Look for an excess of parental types.

Solved Problem 8-11. Corn seeds may be colored *(C)* or white *(c)*; full *(S)* or shrunken *(s)*; and starchy *(W)* or waxy *(w)*. A plant heterozygous for all three loci was produced by crossing two homozygotes, and the resulting heterozygote was test-crossed. Here are the kinds and numbers of offspring:

White, shrunken, starchy	116
White, shrunken, waxy	2
White, full, starchy	626
White, full, waxy	2708
Colored, shrunken, starchy	2538
Colored, shrunken, waxy	601
Colored, full, starchy	4
Colored, full, waxy	113

Are the three loci linked?

Answer. There are 3139 colored, shrunken and 3334 white, full but only 118 white, shrunken and 117 colored, full. That is a definite excess of parental types, so these loci are linked.

There are 2654 shrunken, starchy; 2821 full, waxy; 630 full, starchy; and 603 shrunken, waxy, so again there is an excess of parentals and the loci are linked.

Finally 2542 colored, starchy and 2710 white, waxy are clearly in excess compared with 714 colored, waxy and 742 white, starchy. So this pair of loci also is linked. All three loci are on the same chromosome.

Solved Problem 8-12. Female *Drosophila* heterozygous for three X-linked loci were crossed to males having recessive alleles at all three loci. The following listing shows the male and female offspring separately; the male progeny have a Y-chromosome; female progeny have two alleles for each locus:

RST/rst	1243	*rST/rst*	5
RST/Y	1187	*rST/Y*	8
RSt/rst	926	rSt/rst	9111
RSt/ Y	889	*rSt/Y*	8901
RsT/rst	8820	*rsT/rst*	942
RsT/Y	9215	*rsT/Y*	907
Rst/rst	8	rst/rst	1234
Rst/Y	7	rst/Y	1306

Show that these genes are all linked.

Answer. Each class is made up of both males and females, so add them together in the tabulation. For example, both *Rst/rst* and *Rst/*Y received an *Rst* chromosome from the heterozygous parent in the test cross, so they should be added together in the following computations.

For *R* and *S*, $RS = 1243 + 1187 + 889 + 926 = 4245$
$Rs = 8820 + 9215 + 8 + 7 = 18,050$
$rS = 5 + 8 + 9111 + 8901 = 18,025$
$rs = 942 + 907 + 1234 + 1306 = 4389$
So these loci are linked since parentals are in excess.

For *R* and *T*, $RT = 20,465$
$Rt = 1830$
$rT = 1862$
$rt = 20,552.$
These loci are linked since two classes are in excess.

For *S* and *T*, $ST = 2443$
$St = 19,827$
$sT = 19,884$
$st = 2555$
Again two classes are in excess, so the genes are probably linked.

What is the map order?
You can tell by inspection what the order of the genes are. Suppose you have this map:

A   20 m.u.   B   30 m.u.   C

Consider this heterozygote from a mapping study: *A B c / a b C.* You do a test cross and get these results:

B c A   250   B C A 15   B C a 100   B c a 150

b C a   250   b c a 15   b c A 100   b C a 150

How can we figure out the order of the three loci on their chromosome?

In a three-point test cross, three gene orders are possible. Each order has a different gene in the middle. In this case A B C, A C B and B A C are the three different orders. To figure out the map order by inspection, consider the parental types. You can find them without doing any calculations since they are the most frequent types. In this case, there are 250 each of A B c and a b C.   Now find the classes that are double crossovers.

A crossover can occur between the first two genes or the second two genes.
If so, then there is a possibility that a crossover could occur in both regions simultaneously. Obviously, that will be the least frequent event. In our case, the least frequent classes are BCA and bca.

Consider the parental in each possible map order:

     A B c   A c B   B A c

     a b C   a C b   b a C

Now look at those orders with the double crossovers indicated:

Note that the top strand in the first one was A B c before the double crossover and A b c after. The top strand in the second one was A c B before and A C B after. And in the third case, the B A c order after a double crossover was B a c. In each case, the middle gene switched positions relative to the outside genes.

In our numerical example, the parentals are B c A and b C a, while the doubles are B C A and b c a. The middle gene is C.

PRACTICE PROBLEMS

8-13. Determine the gene order by inspection from the data for Problem 8-11.

8-14. Determine the gene order by inspection from the data for Problem 8-12.

What are the map distances?

From here on, these problems are just like those in the previous section. You've already grouped the pairs and identified which classes are parental and which are recombinant. So now just calculate the percentage of recombinants and you'll have what you need to construct the map.

Solved Problem 8-15. From the data in Problem 8-11, construct a map.

Answer. The distance between the white/color locus and the shrunken/full locus is $(235 \div 6708) \times 100 = 3.5\%$ or 3.5 map units.

The distance between the shrunken/full locus and the starchy/waxy locus is $(1233 \div 6708) \times 100 = 18.4\%$ or 18.4 map units.

The distance between the white/color locus and the starchy/waxy locus is $(1456 \div 6708) \times 100 = 21.7\%$ or 21.7 map units.

By inspection we figured out the map order in Problem 8-13. It is white - shrunken - waxy. You can also figure it out by calculating the map distances. Since 21.7 is the largest map distance, white/color and starchy/waxy are the outside genes, and the map is:

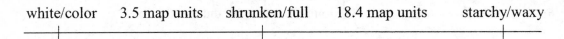

white/color   3.5 map units   shrunken/full   18.4 map units   starchy/waxy

Solved Problem 8-16. Use the data in the answer to Problem 8-12 to construct a map of the *R, T* and *S* loci of *Drosophila*.

Answer. The total number of flies among the progeny is 44,709. For the *R - S* distance, there are 8634 recombinants (*RS* and *rs*), therefore the map distance is $(8634 \div 44709) \times 100\% = 19.3$ map units.

For the *R - T* distance: $(3692 \div 44709) \times 100\% = 8.3$ map units

For the *S - T* distance: $(4998 \div 44709) \times 100\% = 11.2$ map units

The outside loci must be *R* and *S* with the largest distance between them. So the map is:

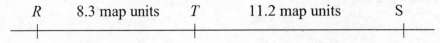

R   8.3 map units   *T*   11.2 map units   *S*

PRACTICE PROBLEMS

8-17. A three-point test cross was done using a plant heterozygous for the genes *M, N* and *P*. One thousand offspring were obtained. The numbers of each phenotype are shown in the following table:

MNP	343	*mNP*	9
MNp	66	*mNp*	61
MnP	79	*mnP*	74
Mnp	11	*mnp*	357

A) Write the symbol for the genotype of the heterozygous parent.
B) If the heterozygous parent was obtained from a cross of two homozygous plants, what were their genotypes?
C) Construct a map of these three genes showing their order and the distance between them.

8-18. Heterozygotes for three loci on the second chromosome in flies were involved in a three-point test cross. From the data on progeny that resulted from this cross, determine the linkage map for these three genes. Figure out the map order by inspection.

DPE/dpe	1951	*DpE/dpe*	40
dpe/dpe	1910	*dPe/dpe*	49
DPe/dpe	179	*Dpe/dpe*	470
dpE/dpe	180	*dPE/dpe*	501

*8-19. Investigators at the Centers for Disease Control in Atlanta isolated three new mutant alleles in the malaria-carrying mosquito *Anopheles gambiae*. One of these new alleles was a recessive that resulted in a difference in body color from wild type mosquitoes. They called the mutation *homochromyl* (symbolized *h*). The wild type allele is *H*. They carried out a three-point test cross involving two other alleles they already had in their collection, resistance to the insecticide Dieldrin D^R (the other allele is sensitivity to Dieldrin D^s) and an eye color allele *c* (the wild type is *C*).

They got these progeny as a result of the test cross:

H	D^R C	359	h	D^R C	6
H	D^R c	9	h	D^R c	292
H	D^s C	274	h	D^s C	12
H	D^s c	4	h	D^s c	367

A) Are any of the loci linked?
B) Make a gene map for the linked loci. Find the map order by inspection. (Benedict, *et al.*, 2002)

CHAPTER 9

CHROMOSOMES

The Chromosome Theory of Heredity puts the genes on the chromosomes.

This book is a step-by-step guide to working genetics problems in the way that Gregor Mendel did, treating the "factors" or genes as abstractions. But the **Chromosome Theory of Heredity** makes it easier by putting the genes on the chromosomes. If you understand the behavior of chromosomes, you should also have a deeper understanding of how genes behave, even beyond Mendel's understanding.

Chromosomes may be dispersed or condensed.

Much of the time in most cells the chromosomes are not visible. At those times, the chromosomal material exists as very dispersed threads of DNA and protein, actively synthesizing RNA and new DNA. In dividing cells, the chromosomal material condenses into tightly packed structures that people can see when the material is stained. These are the chromosomes as they prepare to undergo division during mitosis or meiosis.

Condensed chromosomes have a distinctive anatomy.

> **Chromosomes** in the condensed state have characteristic architecture that sometimes makes it possible to distinguish each chromosome in a cell from every other. Because each chromosome is a little different, they may be cut out from a photograph or drawing and arranged to produce a diagram called a **karyotype.** In a karyotype, chromosomes are arranged in a line in order from longest chromosome to shortest. The number of chromosomes and the structure of each chromosome in a karyotype are characteristics of each species.

Chromosomes may be single or double.

A chromosome consists of one or two **chromatids**. Before division, the chromosomes have two chromatids; after division they have one. But one chromatid or two, a chromosome is still a chromosome. You should think of a chromosome as single after division and double before division, but always just a chromosome. Chromatids may be long or short. The length of chromatids is one way to tell chromosomes apart. (See note on page 118 for further information on chromatids.)

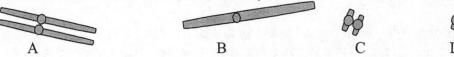

A   B   C   D

101

Chromosome A has two long chromatids.   Chromosome B has one long chromatid. Chromosome C has two short chromatids.   Chromosome D has one short chromatid.

The centromere and the kinetochore result in each chromatid having two arms.

A **centromere** is a small constriction or clear spot along the chromosome. The centromere helps attach a chromosome to the spindle. A **kinetochore** is the part of the centromere that actually attaches to a spindle fiber. In this book, the centromere is symbolized by a little circle. The location of the centromere is one way of telling the chromosomes apart. The centromere may be in the middle of a chromosome or it may be out toward one end. The centromere divides each chromatid into two **arms**. When the centromere is toward the middle, the chromatid has two arms of nearly equal length, but when it is toward one end, one arm is much shorter than the other.

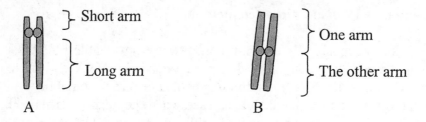

Chromosome A has a centromere toward one end. Its arms are of unequal length. Chromosome B has a centromere toward the middle. Its arms are approximately equal in length.

Other distinguishing features of individual chromosomes include knobs, bands and constrictions.

In some chromosomes, the chromatids are not of equal thickness all along their length. They may have narrow places (constrictions) or wider places (knobs). Often one or more of the chromosomes are attached to a big sphere of RNA called the **nucleolus**. The place where the nucleolus attaches is called the nucleolus organizer.

When chromosomes are stained with certain dyes, they may not stain uniformly. Instead, little bands of dark and light material produce a unique banding pattern on each chromosome. A special stain called the G-banding stain is one way to see patterns. These patterns are another way of telling the chromosomes apart.

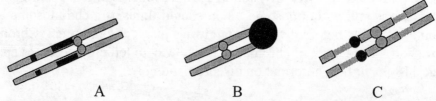

Chromosome A has been stained in a way that brings out a banding pattern unique to this chromosome. Chromosome B has a nucleolus attached. Chromosome C has a series of knobs and constrictions along its length.

Solved Problem 9-1. A) Draw a chromosome with two chromatids, each of which has two arms of about equal length and about equal-sized arms. B) Draw a chromosome with one chromatid, unequal length arm and a knob on the end of the long arm.

Answer. A)   B)

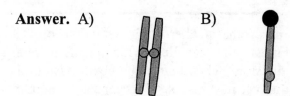

Solved Problem 9-2. Here is a nucleus with the chromosomes condensed. Make a karyotype for this organism.

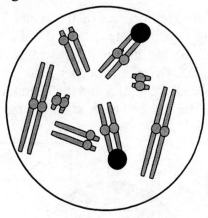

Answer. A karyotype arranges the chromosomes in a linear array from longest to shortest. The easiest way is to make a copy of the picture, cut out the chromosomes and line them up. Note that there are two of each chromosome.

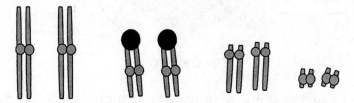

PRACTICE PROBLEMS

*9-3. The black-winged kite, a small hawk-like bird that lives in parts of Africa, Europe and Asia, has 68 chromosomes. The nucleolus organizer is on two very short chromosomes with the centromere nearly at one end. It is typical of birds to have several very short dot-shaped chromosomes. This species only has one such chromosome. Chromosome 1, the largest autosome, has its centromere about 1/4 the distance from one end and shows three small bands evenly spaced along the large arm when stained with their G-banding stain. (See page 102 about G-banding.) Chromosome 5 has two equally long arms, and most of each arm stains with the G-banding stain. Draw Chromosomes 1 and 5. (Bed'Hom *et al.*, 2003)

A key to the cell cycle is mitosis, which keeps the chromosome number constant as cells divide.

Dividing cells go through a cycle of synthesis alternating with division. At the division stage, the chromosomes and most cell organelles divide in half, and half goes to each of the cells that result from division. Then before the next division, the cell synthesizes material to replace what is lost in division. This repeated alternation of division and synthesis is called the **cell cycle.**

Start with cell division. The two processes of the actual cell division are **mitosis**, the division of the chromosomes (abbreviated "M" on the diagram), and **cytokinesis** (abbreviated "C" on the diagram), the division of the whole cell. Once division is over, a period called G_1 occurs until DNA synthesis begins. The period of DNA synthesis is the **S** or **synthesis phase**. Between the end of S and the beginning of mitosis, the cells are in another period called G_2, and then the cell cycle starts over again.

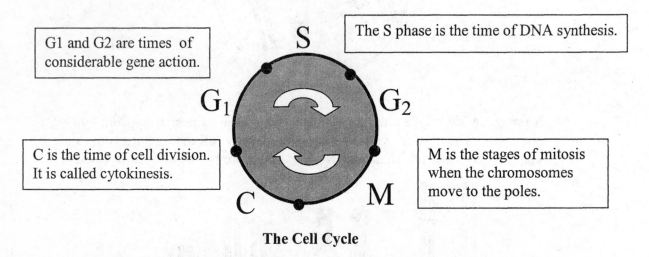

G1 and G2 are times of considerable gene action.

The S phase is the time of DNA synthesis.

C is the time of cell division. It is called cytokinesis.

M is the stages of mitosis when the chromosomes move to the poles.

The Cell Cycle

Mitosis is a set of chromosome movements that assures that each of the two daughter cells receives the same number of chromosomes as the parent cell. Each chromosome starts cell division with two chromatids and ends up after mitosis as a single chromatid. The doubling of the chromosomes to produce two chromatids happens in the S phase.

During the stages of mitosis, chromosomes line up and move apart to separate cells.

The following Figure 1 and Table I summarize the features of mitosis. Table I is a simplification of the process to help you in reviewing your knowledge of chromosomes.

Figure 1

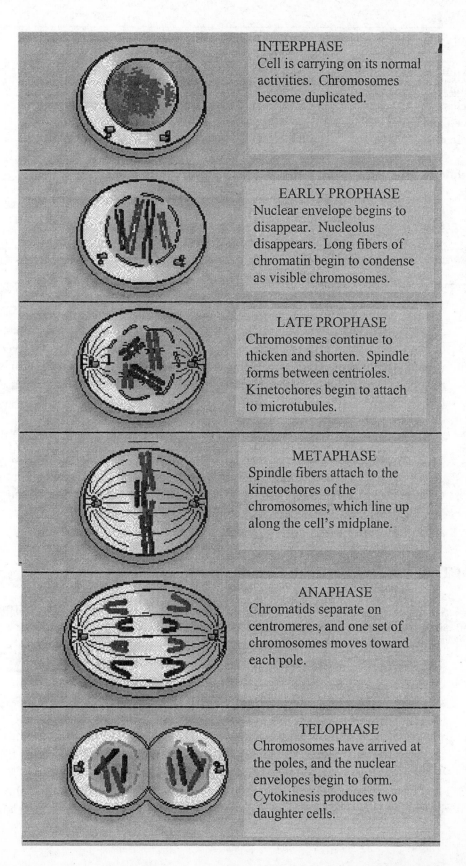

INTERPHASE
Cell is carrying on its normal activities. Chromosomes become duplicated.

EARLY PROPHASE
Nuclear envelope begins to disappear. Nucleolus disappears. Long fibers of chromatin begin to condense as visible chromosomes.

LATE PROPHASE
Chromosomes continue to thicken and shorten. Spindle forms between centrioles. Kinetochores begin to attach to microtubules.

METAPHASE
Spindle fibers attach to the kinetochores of the chromosomes, which line up along the cell's midplane.

ANAPHASE
Chromatids separate on centromeres, and one set of chromosomes moves toward each pole.

TELOPHASE
Chromosomes have arrived at the poles, and the nuclear envelopes begin to form. Cytokinesis produces two daughter cells.

Table I.

	Interphase	Prophase (Early)	Prophase (Late)	Metaphase	Anaphase	Telophase
Chromosome Material	Diffuse	Begins to condense	Condensed	Condensed	Condensed	Diffuse by end
Nuclear Envelope	Intact	Intact	Breaks up	Absent	Absent	Returns
Nucleolus	Intact	Intact	Absent	Absent	Absent	Absent
Spindle	Absent	Absent	Begins to form	Present	Present	Begins to disappear
Chromosomes on Spindle	—	—	—	Lined up on spindle	Moves to poles	Detach from spindle
Chromatids	1 before S 2 after S	2	2	2	1	1

Solved Problem 9-4. Suppose an organism has 16 chromosomes, and each chromosome has arms of approximately equal length. How many chromatids would go to each pole at anaphase? How many arms would you see at late prophase? How many arms and chromatids would you see in a nucleus at telophase before the chromosomes begin to decondense?

Answer. All 16 chromosomes would go to each pole at anaphase. At anaphase, each chromosome has only one chromatid, so 16 chromatids would go to each pole. In late prophase, each of the 16 chromosomes has two chromatids, so you would see 32 chromatids. Each chromatid is divided into two arms, so you would see 64 arms. At telophase each nucleus has half the number of chromatids and arms than at prophase, so you would see 16 chromatids and 32 arms.

PRACTICE PROBLEMS

9-5. A, B and C are individual chromosomes. Which chromosomes are from a cell in anaphase and which are in prophase? How do you know?

A.   B.

C

9–6. In developmental biology, the theory of genome equivalence claims that all the cells in an adult organism have the same genes. How is mitosis evidence for this?

*9-7. The chromosomes of five different frog species from Venezuela and the Caribbean had a diploid number of 24. A sixth species had 22. All six species had six short chromosome pairs and one long chromosome pair, and the rest were intermediate in length.

At anaphase of mitosis, how many intermediate length chromatids go to each pole in the species with a diploid number of 22? How many total chromosomes would we see in prophase cells of the other five species?

*9-8. *Mus musculus* is the common house mouse and laboratory mouse. Like most other species of that genus, *Mus musculus* has a diploid chromosome number of 40. One exception is the Indian spiny mouse *Mus platythrix,* with a diploid number of 26. Chromosomes were prepared from spleen cells grown in a tissue culture medium. What stage of the cell cycle in the spleen cells do you think is probably best for getting good chromosomes? How many centromeres will you see in a karyotype of each of the mouse species? (Matsubara *et al.,* 2003)

Meiosis is the key to sexual reproduction. It reduces the chromosome number in half.

One complete set of chromosomes with one of each chromosome is called a **haploid** set, and that number of chromosomes is the haploid number. Many organisms spend most of their life cycle in a stage in which every cell has two complete haploid sets. Such cells are said to have a **diploid** set of chromosomes, and the number of chromosomes is the diploid number. The members of each pair of chromosomes in a diploid set are called **homologous chromosomes.** In general, homologous chromosomes have the same gene loci at corresponding positions along the chromosomes.

The sexual life cycle. Organisms that reproduce sexually alternate between the production of gametes and the fusion of gametes to form zygotes. The basic sexual life cycle is quite simple, although it is made more complex in some organisms.

The sexual life cycle alternates between fertilization and meiosis.

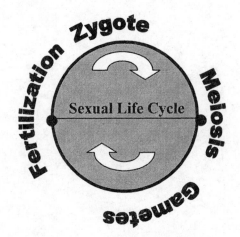

If gametes had the same number of chromosomes as other cells, a problem would arise at fertilization. Each gamete would donate one complete set of chromosomes to the zygote, giving it twice as many chromosomes as the last generation. The function of meiosis is to overcome this problem by reducing the number of chromosomes by half. Not just any half will do either. One complete set of chromosomes must be put in each gamete.

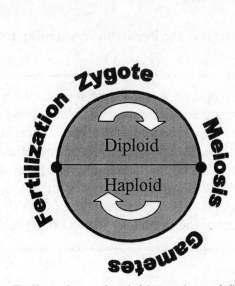

The sexual life cycle also alternates between haploid and diploid.

Solved Problem 9-9. If the forest plant *Trillium* has a haploid number of five, how many chromosomes will be in a karyotype of cells from the diploid stage of its life cycle?

Answer. The diploid number is twice the haploid number, and there is one of each chromosome in the karyotype, so the diploid karyotype will have five pairs or 10 chromosomes.

PRACTICE PROBLEMS

*9-10. *Mus musculus* is the common house mouse and laboratory mouse. Like most other species of that genus, *M. musculus* has a diploid chromosome number of 40. One exception is the Indian spiny mouse *Mus platythrix* with a diploid number of 26. Chromosomes were prepared from spleen cells grown in a tissue culture medium. What stage of the cell cycle in the spleen cells do you think is probably best for getting good chromosomes? How many centromeres will you see in a karyotype of each of the mouse species? (Matsubara *et al.* 2003)

Synapsis is the key to understanding meiosis.

A thorough treatment of meiosis with useful diagrams may be found in any good introductory biology text. There are two especially important concepts to understanding meiosis. First, the chromosomes pair with one another. In a diploid cell there are two of each chromosome, two complete sets. Every chromosome has a homologue. At the beginning of meiosis, each of the chromosomes in one set pairs with its homologue from the other set. They attach to each other very tightly in a process called **synapsis**. Then the synapsed chromosomes line up on the metaphase plate of the spindle in preparation for cell division.

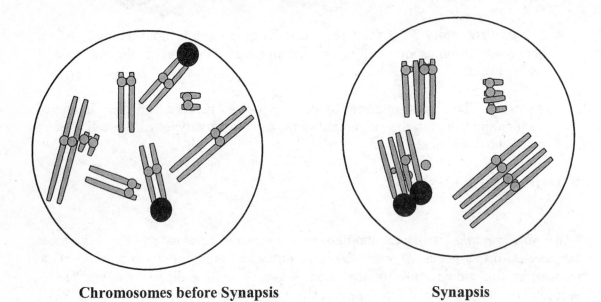

Chromosomes before Synapsis   **Synapsis**

Because each of the chromosomes in synapsis has two chromatids, the pair of synapsed chromosomes has four chromatids and is called a **tetrad**. If you look carefully at a tetrad, you will see that often a chromatid from one chromosome is attached to a chromatid from the other in a characteristic association called a **chiasma**. It appears that the chromatids are exchanging material at the chiasma, and indeed they are. The chiasmata are where crossing over occurs between linked genes.

A Tetrad with One Chiasma

Remember that meiosis includes two cell divisions.

Cells undergoing meiosis divide twice. Each division includes metaphase and anaphase. The first division includes a prophase in which the chromosomes condense and synapse, and the second division includes a telophase. Organisms differ in what happens between divisions, so the first telophase and the second prophase may or may not be missing. The stages of the two meiotic divisions are given either a I or a II to identify them.

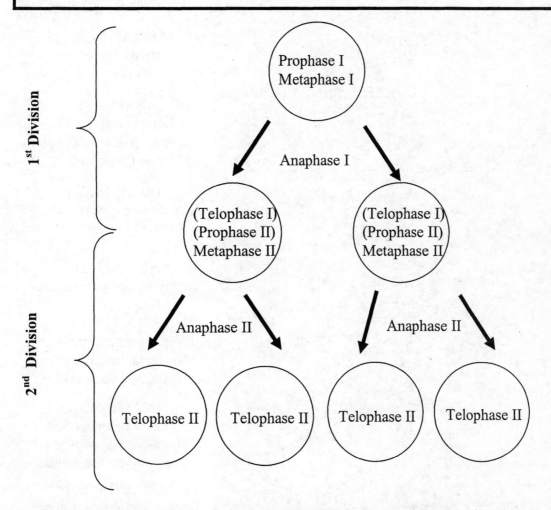

At Metaphase I, the synapsed homologous chromosomes (tetrads) line up on the metaphase plate. At Anaphase I, one of each pair of homologous chromosomes goes to each pole. That means that a haploid set goes to each pole. Each chromosome has two chromatids during this part of meiosis.

At the second division, the chromosomes, which are no longer synapsed, line up at the metaphase plate, and unduplicated chromosomes go to each pole. That means the haploid number is again going to each pole, but this time with the chromosomes unduplicated.

This diagram summarizes the essential features of meiosis.

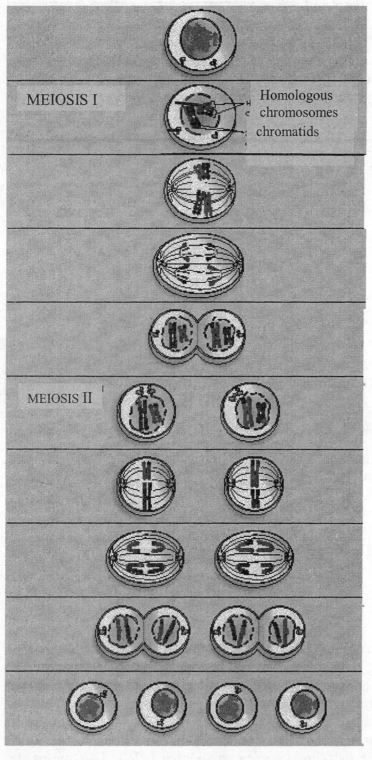

INTERPHASE
Uncondensed chromosomes before meiosis starts.

PROPHASE I
Chromosomes condense, homologous pairs synapse.

METAPHASE I
Tetrads line up on metaphase plate.

ANAPHASE I
Chromosomes move to spindle poles, each with two chromatids.

TELOPHASE I
Not always present. May skip to Prophase II.

PROPHASE II
Ready for second meiosis.

METAPHASE II
Single chromosomes line up on metaphase plate. The haploid number.

ANAPHASE II
Haploid number of chromosomes with one chromatid each to poles.

TELOPHASE II
Nuclear envelopes re-form in haploid cells.

INTERPHASE
Chromosomes once again disperse.

Stage	Synapsis	Chromatids at Anaphase	Chromosome Number Going to Poles	End Result
Mitosis	No	Unduplicated	Diploid	Two cells like the cell that divided
Meiosis I	Yes, in Prophase I	Duplicated (Two chromatids)	Haploid	Two cells, haploid number
Meiosis II	No	Unduplicated	Haploid	Four cells, haploid number

Solved Problem 9-10.   *Homo sapiens* have a diploid number of 46 chromosomes. If we are told that in a cell division being observed, 23 chromatids are going to each pole, can we tell if the cell is in mitosis or meiosis?

Answer. In mitosis, one chromatid of each chromosome goes to each pole. The diploid number is 46 chromosomes, so if we see 23 going to the pole, the division cannot be mitosis. In Meiosis II, half the chromosomes, each with one chromatid, goes to each pole, which in this case would be 23. So we can be fairly sure that we are observing Meiosis II.

Solved Problem 9-11. The diplolid number for a certain species of crab is 100. How many chromatids will go to each pole during Anaphase I? How many chromosomes? How many centromeres?

Answer. The haploid number of chromosomes, which is 50, will go to each pole. Since this is Anaphase I, each chromosome will have two chromatids. So 100 chromatids go to each pole. Each chromatid will have a centromere, so there will be 100 centromeres going to each pole.

PRACTICE PROBLEMS

9-12. "Polyploids" have greater numbers of chromosome sets than a diploid does, sometimes three (triploid), four (tetraploid) or even more. Suppose the haploid set of chromosomes in a series of species with different ploidy levels had one chromosome with a recognizable constriction halfway between the centromere and the end of the long arm. How many such chromatids would you see going to the pole of the spindle in:

A) Anaphase II of a tetraploid?
B) Anaphase I of a diploid
C) Anaphase of mitosis in a hexaploid?

9-13. What problem arises for a triploid or pentaploid organism at meiosis? Why doesn't that problem arise in mitosis?

*9-14. A polyploid has more than one complete set of chromosomes. A triploid has three sets, and a tetraploid has four. An **allopolyploid** has chromosome sets from two or more different species. There is a complete diploid set of chromosomes from each species, so that every chromosome pairs up with a homologous chromosome of its own species at meiosis.

An allopolyploid of two plant species, *Arabidopsis arenosa* and *A. thaliana,* is known by the species name *A. suecica*. A fluorescent dye was made that binds specifically to the centromeres of the two parent species. When you look at the stained chromosomes, the centromeres glow. The probe to *A. arenosa* glowed green and the probe to *A. thaliana* glowed red. Assume that the investigators could see a glowing centromere on every chromatid. They looked at *A. suecica* in late Prophase I of meiosis using their probes, and they found eight sets of four green spots and five sets of four red spots. What is the diploid number of each of the three species? (Comai, Tyagi, and Lysak, 2003)

The behavior of chromosomes explains the behavior of genes in crosses.

If you put alleles on the chromosomes, you can see how the Chromosome Theory of Inheritance explains Mendel's laws as well as other gene behavior.

Mendel's Law of Segregation can be demonstrated with model chromosomes.

Imagine that an organism has a diploid number of four. The chromosomes of that organism are drawn below. Use them to make a model of what happens at mitosis. **Copy this page and cut out the chromosomes. Use these cut-outs as models to follow the progress of meiosis.**

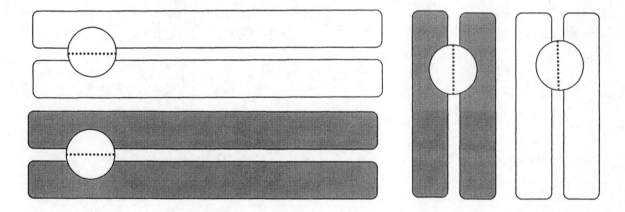

Cut out each chromatid by separating the centromeres along the dotted line. In each case, the gray chromosome came from the father, and the white one came from the mother.

Choose a place on one of these chromosomes and put an allele "A" on both chromatids of one chromosome and an allele "a" on both chromatids of its homologous chromosome.

Then follow the chromosomes through meiosis by arranging them in each of the stages of both cell divisions. By the end of the process, all four cells have an allele of this gene. Two have "A" and two have "a." In a monohybrid cross, this is how a heterozygote's genes behave. Separation of the homologous chromosomes at Anaphase I of meiosis is the basis for Mendelian segregation.

Prophase I of Meiosis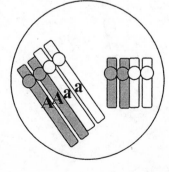

One of Four Possible Gametes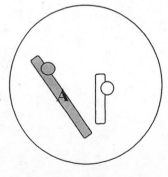

Mendel's Law of Independent Assortment can be demonstrated with model chromosomes too.

> Repeat the demonstration with the addition of a B/b locus on the other chromosome. This time you will have to pay attention to how the chromosomes line up at Metaphase I. That little detail is the basis of independent assortment.

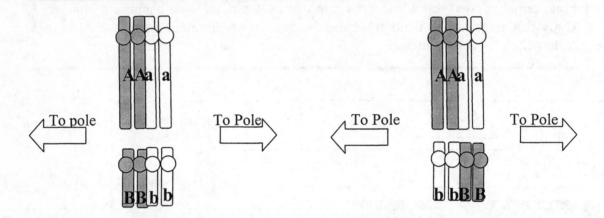

The chromosomes could be this way. If this way, the gametes will end up with an *A* and *B* or an *a* and *b*.

OR

They could be this way. If this way, the gametes will have an *A* and *b* or *a* and *B*.

Solved Problem 9-15. Suppose you were studying an organism like that in Problem 9-2. Describe the gametes if the allele "T" were homozygous.

Answer. The nucleus in Prophase I of meiosis will look like this:

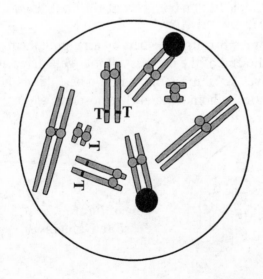

All four meiotic products will be the same and will look like this:

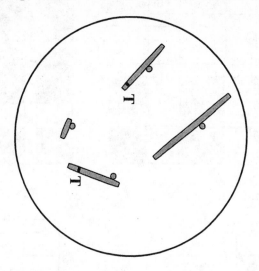

This demonstrates the chromosomal basis for the observation that homozygotes only produce one kind of gamete.

PRACTICE PROBLEMS

9-16. A new technique makes it possible to stain each homologous pair a different color by using stains that bind to gene sequences found only on that one pair. Suppose the haploid set of a fern contained one each of a red-staining, yellow-staining, green-staining, blue-staining and purple-staining chromosome. Ferns have a diploid multicellular stage of the life cycle, and in true plant fashion, a haploid multicellular stage also.

A) How many purple chromosomes will you see in the haploid stage?
B) How many red chromatids will you see in prophase of mitosis in the diploid stage?
C) Which stage of the life cycle undergoes meiosis?
D) Why can only one stage undergo meiosis while both can undergo mitosis?
E) How many different colors will you see in the Telophase II cells at the end of meiosis?
F) Why is the answer to E evidence for the Chromosome Theory of Heredity?

*9-17. In humans, loci for enzymes involved in carbohydrate metabolism have been found on the X-chromosome (phosphorylase kinase and phosphoglycerate kinase) and on Chromosome 11 (glycogen phosphorylase and lactate dehydrogenase). Another gene for carbohydrate metabolism, responsible for enolase, is on Chromosome 17. Use these model chromosomes to show how the chromosomes would be distributed in a male and a female. Assume no crossing over. Then diagram a cross involving these three genes as you learned to do in Chapters 1, 2, 4 and 6. Compare the results of the model chromosome exercise with that of diagramming the cross. How could you modify the model chromosomes to allow crossing over between the linked genes? (Rankinen *et al.,* 2002)

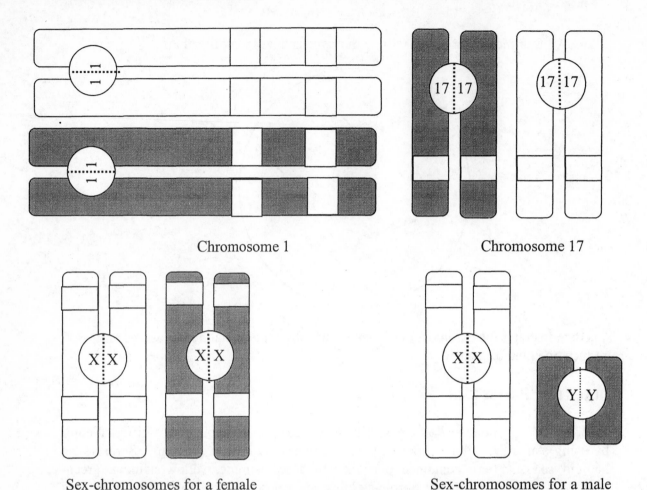

Chromosome 1   Chromosome 17

Sex-chromosomes for a female   Sex-chromosomes for a male

A note about chromosome and chromatid. (See pages 101 and 102.)

Chromatids have been confusing students since students have had to learn about chromatids. This book has taken a simplifying approach to chromatids that helps the beginning student. Further study of genetics will allow the student to tease out the subtleties of what is being learned.

In this book, chromosomes may have one chromatid or two. In either case the structure is a chromosome. With such usage, students can be asked how many chromatids go to each pole at Anaphase I and Anaphase II. This and similar problems helps them absorb the essential point: that one strand or two, a chromosome is always a chromosome.

Many geneticists prefer to say that a chromosome may be duplicated or unduplicated. When duplicated, the chromosome has two chromatids, but the unduplicated chromosome does not have one chromatid. It is simply a single-stranded or unduplicated chromosome. This usage also emphasizes that a chromosome is a chromosome whether with one strand or two. But with this usage, students can't ask how many chromatids go to the poles in Anaphase II and similar problems which help them internalize their understanding of what happens in mitosis and meiosis.

Chapter 10

Population Genetics

Population genetics is the study of gene pools. A **gene pool** is a collection of all the genes in a population. A **population** in this case is all of the individuals of a particular species in a given area. Populations lend themselves to calculating **frequencies** of just about anything that varies in the population, including colors, patterns, size, behavioral characters and, most especially, genes. The ability to calculate frequencies is the first step to understanding population genetics. Most population genetics is done using algebra and calculus and probability studies, most of which is outside the scope of this book.

A frequency is simply the number of individuals of one type relative to the total number in the population. Sample the population and count the number of each type. Then divide the number in a type by the total number in your sample to calculate the frequency. See Chapter 2 to review how similar a frequency is to a probability.

> **Solved Problem 10-1.** An ornithologist noticed that in a flock of wild turkeys, nine had blue heads and four had pink heads. What is the frequency of blue-headed birds?
>
> **Answer.** The total number of turkeys is 13 of which nine had blue heads. The frequency of blue heads is 9/13 = 0.69.
>
> **Solved Problem 10-2.** A man who sold poinsettias had a greenhouse that held 250 plants. He put all of the 50 plants with white leaves on the sunny side of the greenhouse, 75 with pink leaves on the shady side and filled the center with deep red plants. What is the frequency of each of the color forms in this collection?
>
> **Answer.** The pink and white together are 125 of 250. Thus, the red are 250 – 125 = 125. White frequency is 50/250 = 0.2. Pink frequency is 75/250 = 0.3. Red frequency is 125/250 = 0.5.
>
> **Solved Problem 10-3.** I liked 15 of the paintings in the new art show, but I could have done without the other three. What was the frequency of paintings I liked?
>
> **Answer.** 15/18 = 0.83

Generalizations about frequencies:

1. They will always have a value from 0 to 1. A frequency of 1 means all the objects are alike for that population. A frequency of 0 means no individuals of that type are in the population.
2. The total of all frequencies will be 1.
3. We symbolize frequencies in this way: The symbol $f_{(A)}$ means "the frequency of A" where A is one class of a character being studied. So in Problem 10-2, we could write $f_{(white)}$, $f_{(pink)}$ and $f_{(red)}$.

4. For thinking generally, we can let one frequency of A be p and the frequency of a be q. $p + q = 1$. For three categories, $f_{(A)} = p$, $f_{(B)} = q$ and $f_{(C)} = r$, with $p + q + r = 1$.

Gamete frequencies to zygote frequencies.

Population geneticists think of the gene pool as a collection of gametes from which zygotes are made at fertilization. If we know the frequencies of all the gamete types, we can calculate the zygote frequencies.

Suppose a locus of a diploid organism has two alleles A and a. Let $f_{(A)} = p$ and $f_{(a)} = q$. Since these are the only alleles, $p + q = 1$. Now combine eggs and sperm in all possible combinations. We can use a Punnett square to keep track of the frequencies of each gamete type. Usually the frequencies of eggs are put along the side of the Punnett square, and frequencies of sperm along the top.

	A allele from sperm $f_{(A)} = p$	a allele from sperm $f_{(a)} = q$
A allele from egg $f_{(A)} = p$	AA zygote $f_{(AA)} = p \times p = p^2$	Aa zygote $f_{(Aa)} = p \times q = pq$
a allele from egg $f_{(a)} = q$	Aa zygote $f_{(Aa)} = p \times q = pq$	aa zygote $f_{(aa)} = q \times q = q^2$

Solved Problem 10-4. If we were to take out a large number of eggs and sperm from the gene pool and let the gametes combine at random, what would be the genotypes of zygotes and their frequency? Let $f_{(A)} = 0.8$ and $f_{(a)} = 0.2$ in this gene pool.

Answer: We can use the Punnett square once again but this time using specific frequencies of A and a. The frequency of AA is $p^2 = 0.8^2 = 0.64$. The frequency of Aa is $2pq = 2 (0.8)(0.2) = 0.32$, and the frequency of aa is $q^2 = 0.04$.

	A allele from sperm $f(A) = p = 0.8$	a allele from sperm $f(a) = q = 0.2$
A allele from egg $F_{(A)} = 0.8$	AA zygote $0.8 \times 0.8 = 0.64$	Aa zygote $0.8 \times 0.2 = 0.16$
a allele from egg $f(a) = q = 0.2$	Aa zygote $0.8 \times 0.2 = 0.16$	aa zygote $0.2 \times 0.2 = 0.04$

A Little Help from the Expansion of Binomials

The expression p + q = 1 is called a binomial. A binomial can be "expanded" by squaring it, that is, by multiplying the binomial (p + q) = 1 by itself $(p + q) \times (p + q) = p^2 + 2pq + q^2 = 1$. Note that the population genetic way of expressing the frequencies of gametes is a binomial and that expanding (p + q) = 1 gives us all possible combinations of gametes just as a grid does. These properties of the binomial make it easier to determine the frequencies of gametes more readily than with the grid. More on binomials and how to expand them is in Chapter 3. We need not worry about expanding a binomial beyond $(p + q)^2 = 1$ because organisms usually combine gametes in pairs.

When p + q = 1 is expanded to $p^2 + 2pq + q^2 = 1$, p is the frequency of A, q is the frequency of a, p^2 is the frequency of AA, $2pq$ is the frequency of Aa, and q^2 is the frequency of aa.

Solved Problem 10-5. Blood groups are useful for study of populations for several reasons. Some blood group alleles lack dominance, so we can unambiguously assign heterozygous and homozygous status to each individual and thus figure out genotypes directly. For each of these blood group data, predict the zygote frequencies:

Kidd blood group: a allele frequency = 0.4358
                      b allele frequency = 0.5142
Auberger blood group: a allele frequency = 0.6213
                      b allele frequency = 0.3787

Answer: Kidd: $p = f_{(a)} = 0.4358$ $q = f_{(b)} = 0.5142$ $(p + q)^2 = 1$
$p^2 + 2pq + q^2 = 1$ $f_{(aa)} = p^2 = (0.4358)^2 = 0.1899$
$\qquad\qquad f_{(ab)} = 2pq = 2(.4358)(0.5142) = 0.4482$
$\qquad\qquad f_{(bb)} = q^2 = (0.5142)^2 = 0.2644$

Auberger $p = f_{(a)} = 0.6213$ $q = f_{(b)} = 0.3787$
$p2 + 2pq + q2 = 1$ $f_{(aa)} = p^2 = (0.6213)^2 = 0.3860$
$\qquad\qquad f_{(ab)} = 2pq = 2(0.613)(0.3717) = 0.4785$
$\qquad\qquad f_{(bb)} = q^2 = (.3787)^2 = 0.1434$

Zygote frequencies from gamete frequencies.

Note that zygote frequencies can be determined directly from gamete frequencies, and gamete frequencies can be determined directly from zygote frequencies. This is an inevitable consequence of the life cycles of plants and animals. In diploid animals, organisms alternate between meiosis and fertilization to produce gametes and zygotes respectively. Plants add a step, producing spores by meiosis and then gametes from a gametophyte generation. This is followed by meiosis in the sporophyte to produce spores. (See Chapter 9.)

Solved Problem 10-6. How are gamete frequencies determined from zygote frequencies? Consider this question first for the general population in which the gamete frequencies are p^2, $2pq$ and q^2, and then for a specific case in which zygote frequencies were AA 0.64, $A\,a$ 0.32 and aa 0.04. What are the gamete frequencies?

AA produces all *A* gametes.

Half of the gametes produced by *Aa* are also *A*.

So in our example $f_{(A)} = f_{(AA)} + 1/2 f_{(Aa)} = p^2 + pq = 0.64 + 0.16 = 0.8$.

By similar reasoning,

aa produces only *a* gametes.

Half the gametes produced by *Aa* are also *a*.

$F_{(a)}$ is $q^2 + pq = 0.04 + 0.16$.

Thus $f_{(a)} = 0.04 + 0.16 = 0.2$.

Practice Problems

10-7. A field biologist found albinos in a population of 500 gophers that she was studying in southern Wyoming. She knows that the albino allele is recessive (*c*) while the normally pigmented allele is dominant (*C*). What will be the frequency of albinos in this population if 481 are not albinos?

10-8. This question is based on the same population as in Problem 10-6. Could the biologist also find the frequencies $f_{(AA)}$, $f_{(Aa)}$ or $f_{(a)}$?

10-9. Suppose she (the woman studying gophers in Problem 10-7) could tell heterozygous gophers from homozygous ones, and her 500 gophers included heterozygotes at a frequency of 0.396. What then will be the gamete frequencies and zygote frequencies at her study site?

10-10. For each of the three populations, predict what the frequencies of gamete types and zygote types are in the next two generations. Assume that you can tell both homozygotes from the heterozygote. Start with calculating the gametes.

Population 1 $f_{(AA)} = 0.5$ $f_{(Aa)} = 0.2$ $f_{(aa)} = 0.3$

Population 2 $f_{(AA)} = 0.25$ $f_{(Aa)} = 0.5$ $f_{(aa)} = 0.25$

Population 3 $f_{(AA)} = 0.04$ $f_{(Aa)} = 0.6$ $f_{(aa)} = 0.36$

A Mathematical Model: The Hardy-Weinberg Equilibrium

Scientists often work with **mathematical models.** A model is a simplification of a system that makes it easier to study. Mathematical techniques are used for further simplicity and also to make the conclusions more general. A model tests assumptions by comparing the behavior of the model with nature.

Often evolution is defined as a change in gene frequencies over time. Population geneticists have found five things that can change the frequencies in a gene pool: mutation, migration, selection, random mating and small sample size.

1. **Mutation** is an inherited change at a locus. In our model, two mutations are possible: A could mutate to a increasing q and decreasing p. Also a could mutate to A, raising p and decreasing q.

2. **Migration** is the movement of members of the same species from one population to another. If individuals are migrating out of a population they are said to be carrying on emigration. If individuals are migrating into a population, they are carrying on immigration. If p and q are different in two populations, migration can change the allele frequencies.

3. **Selection** is the differential success of the reproduction of genes. Phenotypes associated with some genotypes make it more probable that they will pass on their genes to the next generation. When selection is happening, some genotypes will increase at the expense of others. This is how adaptive variation arises in populations. While any of these five forces can cause a change in gene frequency, only selection causes adaptive change.

4. **Random mating** is the choice of mates without regard to the frequency of characters in the gene pool. Inbreeding is non-random mating preferentially with individuals more closely related to oneself than average. Out breeding is mating preferentially with individuals less like oneself than average.

5. A **small sample size** results in fluctuations of frequency due to a random sampling error.

The Hardy-Weinberg Equilibrium makes the rather obvious observation that if none of these forces is acting to change the frequencies of alleles in a gene pool, the frequencies of alleles will not change.

The Hardy-Weinberg Equilibrium is a mathematical model about changes in frequency in populations. The model simplifies a population to one or a few loci, each with two or more alleles. Equilibrium values of the alleles can be calculated and the results are compared to wild populations to see if any of several forces is causing change in the allele frequencies.

The population being studied is not in Hardy-Weinberg Equilibrium if any of the five forces is acting. If no force is acting, the population gamete and zygote frequencies will remain the same in every generation, cycling between $(p + q) = 1$ and $(p^2 + 2pq + q^2) = 1$. Often when one or more of the five forces is working more is acting, the frequencies of the alleles in the population will usually not be in Hardy-Weinberg Equilibrium. But if a population not in equilibrium is returned to Hardy-Weinberg conditions, the frequencies will go to equilibrium values in one generation.

Solved Problem 10-11. Calculate the frequency of alleles A and a in each of these populations. Then determine the frequency of zygotes produced from a gene pool with those allele frequencies.

	AA	Aa	aa	A	a	AA	Aa	aa
Population 1.	0.7	0.0	0.3	___	___	___	___	___
Population 2	0.6	0.2	0.2	___	___	___	___	___
Population 3	0.49	0.42	0.09	___	___	___	___	___

Answer.

	AA	Aa	aa	A	a	AA	Aa	aa
Population 1.	0.7	0.0	0.3	0.7	0.3	0.49	0.42	0.09
Population 2	0.6	0.2	0.2	0.7	0.3	0.49	0.42	0.09
Population 3	0.49	0.42	0.09	0.7	0.3	0.49	0.42	0.09

Note that although each population had different zygotic frequencies, they all produced the same gamete frequencies. If you use these gamete frequencies for all three populations, they will, of course, all have the same zygote frequencies. Thus, in this we can see two phenomena: Several different zygote frequencies can generate the same gamete frequencies. Some frequencies keep the population constant when the next zygotes are produced. Some are not.

That set of gamete frequencies is identical in this case to that of Population 3. This brings us to the main point in this chapter, the Hardy-Weinberg Equilibrium.

Here is how to decide if a population is in Hardy-Weinberg Equilibrium:

1. Begin by determining the zygote frequencies.
2. Use the numbers you get for the zygote frequencies to determine the gamete frequencies.
3. Use those gamete frequencies to determine the next set of zygote frequencies.
4. If the second set is significantly different from the first set, the population is probably not in equilibrium.

PRACTICE PROBLEMS

10-12. A large population of squirrels, many with black fur, live in and around the city park. The black ones are all recessive homozygotes, including the five he captured. The 20 heterozygotes and the 25 homozygous gray squirrels complete his collection. Is the squirrel population in Hardy-Weinberg Equilibrium?

10-13. Several examples of population frequencies are in the problems of this chapter. For each of those populations listed, decide if the population is in Hardy-Weinberg Equilibrium.

Population from Problem	p^2	2pq	q^2	p	q	Equilibrium?
10-4	0.64	0.32	0.04	0.8	0.2	
10-5 (Kidd)	0.1899	0.4482	0.2644	0.4358	0.5142	
10-10 1	0.5	0.2	0.3			
10-10 2	0.25	0.5	0.25			
10-10 3	0.04	0.6	0.36			

Hardy-Weinberg and Multiple Alleles

A population may have more than two alleles in its gene pool. For example, *Drosophila* has many alleles at the white eye locus, from brick red to white, and the ABO blood groups are produced by the action of three alleles, A, B and O.

You can deal with a case of multiple alleles (see Chapter 4) using a modification of the number of variables. How would you find the zygote types if there were three alleles in a diploid population?

Call the alleles *a, b,* and *c* and designate the symbols for the frequency of the alleles to be p, q and r respectively. The three alleles are all the alleles there are, $(p + q + r) = 1$. You can expand this function by multiplying it by itself. To do that, every term in one parent must be multiplied by every term in the other, a task made simpler by using a grid:

	P	q	r
p	p^2	pq	pr
q	Pq	q^2	qr
r	Pr	qr	r^2

$$(p + q + r) = 1 \longrightarrow (p + q + r)^2 = 1 \longrightarrow (p^2 + q^2 + r^2 + 2pq + 2pr + 2qr) = 1$$

By analogy with two-allele problems,

Two-allele problem:
$$(p + q)^2 = 1$$

$$(p^2 + 2pq + q^2)^2 = 1$$
$F_{(aa)} = p^2$
$F_{(bb)} = q^2$
$F_{(ab)} = 2pq$

Three-allele problem:
$$(p + q + r)^2 = 1$$

$$(p^2 + q^2 + r^2 + 2pq + 2pr + 2rq) = 1)$$
$F_{(aa)} = p^2$
$F_{(bb)} = q^2$
$F_{(cc)} = r^2$
$F_{(ab)} = 2pq$
$F_{(ac)} = 2pr$
$F_{(bc)} = 2qr$

Practice Problems

10-14. Expand each of these binomial-like expressions by squaring them.

A) $(p + q)^2 = 1$ B) $(p + q + r)^2 = 1$ C) $(p + q + r + s)^2 = 1$

10-15. A population had a locus with four alleles. The alleles and their frequencies were v (0.1), w (0.4), y (0.2) and z (0.3). What are all the zygote frequencies this population will produce from these gametes?

10-16. This question is about the same population in Problem 10-15. Show the next generation of gametes and decide if the population is in Hardy-Weinberg Equilibrium.

CHAPTER 11

QUANTITATIVE GENETICS

Careful observation of nature provides scientists with data. Numerical data are preferred, and our analysis of this information uses various **descriptive** and **inferential statistics.** Scientists point out that there are at least five categories of data -- each with very different properties -- and we must be sure we know which one to use in any given case. Thus, statisticians warn that to take advantage of all the features of the various statistical tests, a plan for analyzing the data should be part of the overall experimental design.

Although detailed descriptions of many statistical tests are not part of this book, there are some statistical ideas discussed that are necessary in understanding the genetics topics in this chapter.

> **Quantitative genetics** requires statistical tools not used in other fields of genetics. Mendel set the norm for many geneticists by classifying characters on a nominal scale: tall and short plants, apical and basal budding, round and wrinkled seeds. Quantitative traits are more difficult to classify and require simple descriptive statistics to do so.
>
> The following sections describe how to compute each statistic we use from formulas. The various descriptive statistics can be obtained at the click of a button on your computer if you know how to use computer spreadsheets or graphing calculators.

Mendel studied so-called **nominal** or **discontinuous** data. You are using the nominal scale if you are asking, "Which category does this individual belong to?" The data are called discontinuous because there are gaps between the classes of data. You can use the Chi-square test described in Chapter 3 with nominal data.

Quite different data are called **interval, ratio** or **continuous**. To the statisticians these three terms differ slightly in meaning, but many treat them all as one type, as will this book. Using whatever measuring instrument is appropriate – a balance for mass, a tape measure for distance, a spectrophotometer for an action spectrum, a graduated cylinder for volume – we measure as many as is practical.

Consider some data from the beginnings of genetics on domestic animals and plants, published by E. Davenport in 1909. There may be other, more up-to-date uses for these data, but the data remains the data, whatever explanation we want to apply.

Length	Number
5	1
5.5	4
6	6
6.5	7
7	19
7.5	31
8	37
8.5	59
9	46
9.5	39
10	23
10.5	11
11	2
11.5	1

Each number in the first column is a measurement in inches of ear length of a strain of corn. The second column is the number of ears having that length. For example, 6 ears had a length of 6, and 23 had an ear length of 10. Such a data table is often called a "frequency distribution."

Solved Problem 11-1. What can we learn from this data table?

Answer: The least number of individuals is in the classes on the end of the distribution. The most common ear lengths are in the middle of the distribution. There is as much as a 5-inch difference between the length of the longest ear and the shortest one. There is considerable variation in this trait, and the variability needs to be accounted for. It would help if we had mathematical expressions to tell us the magnitude of the differences.

Population and Sample. It is usually not possible to measure a characteristic of interest in every member of a human population. The number of individuals is too great, and they may hide or live in a habitat that is difficult to reach or inhospitable to humans. Therefore, it is necessary to study a sample that is representative of the population. How should we sample so as to avoid bias? How large should the sample be? Should we measure the characteristic being studied in the lab or in the organism's natural environment? All of these questions are part of what is called **experimental design**, a skill that scientists spend much of their time developing.

There are widely used mathematical techniques developed to determine the central tendencies of populations and the amount of spread there is in the variability that is found in most populations. Take a look at several measurements of central tendency and spread.

Descriptive Statistics Commonly Used in Quantitative Genetics

Measures of Central Tendency

It is common when looking at a frequency distribution to wonder what the typical value in a distribution is and what value summarizes all of the values collected. One commonly held idea is that the most representative data point ought to be somewhere in the middle. Several techniques for finding a middle value exist in regular practice by biologists.

Mode

One of the simplest to determine is the **mode**. It is the value with the most individuals on the distribution. This technique requires sorting the data into categories, and the category with the most individuals is called the mode.

The mode is not very helpful if the distribution has two or more maxima, a **bimodal distribution.** It isn't obvious how one can apply the same math to a bimodal distribution as to a regular **unimodal** one.

Mean

A more complex and widely used value of central tendency is the **mean**, also called the **average**. To calculate the mean, find the sample size, or the number of individuals in the sample and the sum of all the individuals in the population (n). Divide the total by n. An equation describes the mean where $\bar{x}$ is the mean, $\sum x$ is the sum of all the values and n is the sample size: $\bar{x} = \dfrac{\sum x}{n}$.

The mean is a good measure of central tendency as long as there are few extreme values. The mean is sensitive to extreme values, and for that reason is not so good for small sample sizes.

Solved Problem 11-2. Determine the mean and the mode of the frequency distribution of corn ear lengths.

Answer. Observe the numbers in the first column and the numbers of ears in the second column of the table. The largest number of ears is in the 8.5-inch category in which there are 59 ears. So the mode is 8.5.

Ear length	Number	Length X number
5	1	5
5.5	4	22
6	6	36
6.5	7	45.5
7	19	133
7.5	31	232.5
8	37	296
8.5	59	501.5
9	46	414
9.5	39	370.5
10	23	230
10.5	11	115.5
11	2	22
11.5	1	11.5
	Sum = 286	Sum =2435

Most ear-length classes contain more than one ear, so that adds an additional step to calculating the mean. It is necessary to multiply each value (ear length) by the number of ears of that size in order to include every ear in the calculation. We can add a column to our table to make it easier to keep the data organized. The sum of all the values of (length X number) is the sum we need to calculate. The sample size is the sum of the middle column. Now use the equation:

$$\bar{x} = \frac{\sum x}{n} = 2435/286 = 8.514.$$

8.514 is the mean.

Median

This last measure of central tendency is not so sensitive to extreme values and so is especially conducive to small sample sizes. To determine the **median**, list the values from highest to lowest in numerical order. Then find the value right in the middle with half the values above it and half below. This is the median.

If you have an odd number of samples, the median will have an equal number of values on each side of one of the values; but if you have an even sample size, there will be no actual representative of the median score. You must add the two values that are on either side of the halfway mark and divide by two.

The median

Odd sample size: ...... 140 145 (150) 154 158 ....
Even sample size: ...... 140 145 150 154 158 ....

Halfway: The median = (145 + 150)/2 = 295/2 = 147.5.5

Now consider the ear-length data given earlier and find the median. There are 286 individuals; you don't have to line up every one of the 286 in numerical order. That's already been done in putting the data into its data table. The median will be in the middle of this, and the data sheet lists the numbers in order of size. The exact half of the distribution is at 286/2 = 143. So start at the left end of the distribution and add individuals as you go until you have added 143 individuals.

Start counting on the left end of the row, keeping a running total: 1 + 4 = 5; 5 + 6 = 11; 11 + 7 = 18; 18 + 19 = 37; 37 + 31 = 68; 68 + 59 = 127.

The next 46 individuals all have 9-inch ears and will carry us past the midpoint to 173. So the median is between two ears of 9 inches in length. The median is 9 inches.

Solved Problem 11-3. Determine the median value of the corn ear lengths used in the previous problem. Then add one additional ear, 5.5-inches long to its grouping. There will be 5 in that class now instead of 4. Calculate the mean and the median for this new data table.

Answer: We don't have to consider every number in the data tables because a very small alteration has been added to one bin. The second bin now has a number of 5. We multiply 5 by 5.5, a new value of 29.5 in the third column instead of 22.

For the mean: Change the value in the middle column from 4 to 5 and in the third column change the second bin from 22 to 29.5. That results in a new sum in the middle column of 291.5, and a new sum in the third column of 2440.5. We have a new mean of 2477/293 = 8.45. The original population had as mean of 8.514. So adding just one ear to bin 2 cut nearly 0.1 inches.

For the median: To determine the effect of the same change on the median, we add an ear in the second bin of the distribution. That will shift every ear after that one ear up the scale from before addition of that one ear. That moves the median up one step, but that is still in the 9-inch ears. So the median is still 9 inches.

PRACTICE PROBLEMS

11-4. *Lumbriculus variegatus* is an annelid worm, which, like others of the phylum, is composed of a string of segments. As part of a larger program of research on regeneration in *Lumbriculus,* students cut worms into random lengths then counted the segments in each length. The biologists doing this study soon realized that it was not an experiment they could do much with in terms of research, but at least it could be used as a problem for learning to calculate the central tendencies: mean, median and mode.

Here are the data in number of segments: 9, 12, 8, 6, 10, 7, 13, 4, 2, 11.

Determine the mean, the median and the mode of these numbers.

11-5. Suppose that three more lengths measuring 9, 10 and 11 segments long were prepared and added to the lengths described in Problem 11-4. Use a data table to help calculate the median, the mode and the mean.

11-6. A group of statisticians received bags of chocolate cookies for dessert. Instead of eating the cookies, they took them apart and counted the number of chocolate chips in each cookie. Here are their numbers in no particular order. Make a data table of these numbers: 7 4 1 3 0 5 3 1 6 4 2 2 4 4 5 4 3 5 6 0 5 7 4 2 3 4 3 2 5 5 6 2 0 3 3 4.

Then calculate the mean, the median and the mode.

11-7. A professor has five students in his seminar. He totaled up their scores at the end of the year and found these data. What is the mean and the median for each student?

Student	Test 1	Test 2	Test 3	Test 4	Mean	Median
Irwin	79	80	85	84		
Lola	65	84	70	78		
Merry	90	95	89	91		
Wilson	95	84	89	64		
Meng	83	70	70	84		

Measures of Variability

When you look at most characteristics long enough, the variabilities in the traits are inescapable. Seldom, if ever, are all the values exactly like the mean. The very existence of a data table, such as the one of corn ear length described in the last section, is an example of **variability**. Many methods have been developed to give a mathematical account of variability:

Range

A very approximate measure is provided by the **range**. It is simply the difference between the highest and lowest values in the frequency distribution. Generally, this measure of variability is at the mercy of the occasional strong outlying data point. On our ears of corn data chart the mean is 8.514; the range of lengths is 11.5 – 5.0 = 6.5.

The range is one of those rough-and-ready estimates of variability, but it has limited usefulness. For example, one drawback is that it does not take sample size into account. If there were a sample size of 28, a range of 16.5 would seem to be a greater variability than the same range in a population size of 286.

Variance

A widely used measure that does take the sample size into account is the **variance**. It also takes all the individual deviations of each individual in the distribution into account.

Consider the corn ears data that we used in the section on central tendency. The variance is found by subtracting the mean of the population from each individual, then squaring the result and adding all these numbers together followed by dividing by n-1 (one less than the sample size).

Ear length	Number in each category	Number – mean	(Number – mean)^2	(Number – Mean)^2 X number
5	1	- 3.51	12.32	12.32
5.5	4	-3.01	9.06	36.24
6	6	- 2.51	6.30	37.80
6.5	7	- 2.01	4.04	28.28
7	19	-1.51	2.28	43.32
7.5	31	--1.01	1.02	31.62
8	37	0.51	1.06	9.62
8.5	59	-0.01	0	0
9	46	9.5	0.24	11.04
	39	0.99	0.98	38.22
10	23	1.49	2.22	51.06
10.5	11	1.99	3.96	43.56
11	2	2.49	6.20	12.40
11.5	1	2.99	8.94	8.94
	Sum =sample size = 286			Sum = 364.41

The formula for the variance is $s^2 = \dfrac{\sum (x - \bar{x})^2}{n-1}$. Refer to the table above as you read the step-by-step description of how to calculate variance.

1. Put the categories in Column 1 into the table in numerical order. Column 2 indicates how many specimens are in each group. As you saw in the section on the mean, this is important information, but don't enter it yet.

2. Subtract the mean of the characteristic being studied from each value measured. In the case of this data sheet, the mean was 8. 514, so subtract 8.451 from each value in Column 1 and enter the result in Column 3. For instance, the number in the first row of column 1 is 5.0. When 8.451 is subtracted from 5.0, it results in -3.51.

3. Square the results and enter that result into the Column 4.

4. Add up all the resulting numbers and you have the sum.

5. Now it's time to obtain the sample size by using the Column 2. This is the number of ears which were counted to get the frequency distribution.

6. Divide the number of ears in Column 2 by the sample size minus one, and you get the variance.

Standard Deviation

The **standard deviation** is the square root of the variance. The variance is usually calculated in order to get the standard deviation. The standard deviation is the most frequently used measurement in thinking about variability of data. In a large sample from a population, about 69% of the values are within ± 1 standard deviation of the mean, and just more than 93% lay within two standard deviation. For the ear length data, symbolize variance as s^2 = variance = 1.274. The standard deviation is the square root of the variance s = 1.13.

Coefficient of Variation

If you want to compare the variability of two different characteristic the standard deviation cannot do the job. For instance, if you want to compare the amount of variability in corn ears to the variability of root weight, the standard deviation is in units of weight in the one case and in units of length for the other.

A better comparison is the **coefficient of variation**, which is found by dividing the standard deviation by the mean of the same measurement. The result is a number without units that lends itself to comparison. The statistic is easy to use if you have calculated the mean and standard deviation.

PRACTICE PROBLEMS

11-8. It is unlikely that most investigators would work with a sample size as low as 5, but it makes for simple exercises for working with descriptive statistics. Determine the mean, median, variance, standard deviation and coefficient variation of these "miniature" problems:

A. 1, 2, 1, 3, 1   B. 5, 6, 7, 8, 9   C. 60, 30, 25, 55, 45

11-9. Here is a problem in which the data are skewed toward the large end of the distribution. A colonial web-building spider spins its web next to other spiders of the same species. If a prey hits a web and is stuck, that web's spider captures the prey and wraps it in silk for later use. If the prey bounces off or rips itself free, it often is caught by another web, and then the second web's spider does the capture. These data show the success of catching prey as a function of the size of the prey. Calculate the mean, median, mode, variance and standard deviation.

1	2	3	4	5	6	7	8	9	10	Prey size (mm)
34	24	18	12	6	4	3	2	1	0	Number caught

ode

11-10. Why must you have at least two values to compute a standard deviation?

GRAPHS AND QUANTITATIVE GENETICS

A **graph** allows the relationships between variables to be more readily understood. For example, when analyzing data, graphs help the investigator discover aspects of the data that might have been missed in a table of numbers. Also, when reporting on research to others, graphs can make certain points about the data easier to see.

The Histogram

The type of graph most vital to the study of quantitative genetics is the **histogram**. This is a bar graph in which its X-axis is a value, such as height or weight or resistance to a pesticide, and the Y-axis shows how many individuals have that value.

Below is a histogram for the ear-length data used as an example throughout this chapter. The data table is to the right of the graph, and the step-by-step description of the construction of corresponding graph is explained below.

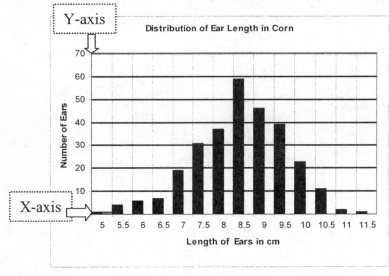

Length	Number
5	1
5.5	4
6	6
6.5	7
7	19
7.5	31
8	37
8.5	59
9	46
9.5	39
10	23
10.5	11
11	2
11.5	1

The Axes

Make X- and Y-axes appropriate for the data. When these data were collected, they were not in the neat categories of the data table. Possibly there was a truckload of corn on its way to becoming frozen corn, and a sample of ears was taken out at the loading dock at the cannery to determine the quality of the truckload. Then when the measuring was done, they had to decide how many categories to use. Obviously, three categories would be too few. Too much detail would be lost in a big category like 7 – 9, for too much of the detail of variability would be lost. See the next pair of graphs. One has the intervals along the X every 3 cm, and the other, every 1 cm.

Plotting of each ear's length separately would spread out the variability into possibly meaningless differences seen in small samples due to high sampling error. A grouping is desired that is large enough to reveal the patterns in the data. This has to be decided on a case-by-case basis. Often it is necessary to make graphs with a couple of different groupings to see what is best.

When gathering the data into groups, each group may contain several different measures depending on how many samples go into each group. When the size of the categories is set, we assume that the objects in each category are equal in size to the value of the midpoint of the category. Sooner or later you will come up with a data point that is right on the line between two categories. A common practice is to decide ahead of time how to count values on the border of two groups. Usually most people agree to put all the borderline data in either the category below the line or above it.

The histogram is not the only form data can take in a frequency distribution. A line graph that connects the Y-value at each X point (as shown in the graph below, to the right of the histogram) sometimes is helpful in visualizing the data. Compare these two forms below, then make these two types of graphs for the data of Problem 11-9.

For quantitative traits, the mean and standard deviation may be thought of as the phenotype for these traits. The normal curve portrays both – the mean is proportional to the height and the width to the standard deviation. Look at the ideal normal curve below. The height of the center vertical line on each is the mean, and the other vertical lines are the boundaries of the first and second standard deviation. Note that the two descriptive statistics need not vary together – tall means and short are either one found with wide standard deviations or narrow.

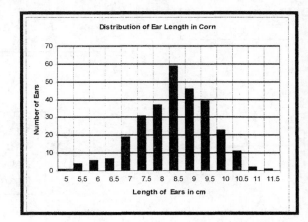

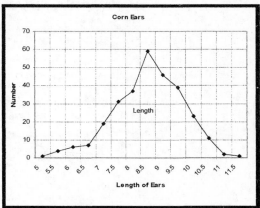

Ideal Normal Curves with different standard deviations.

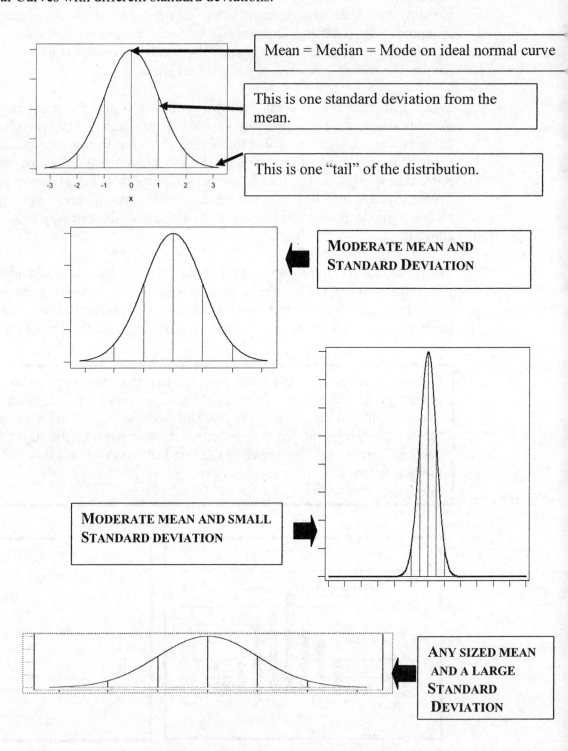

Mean = Median = Mode on ideal normal curve

This is one standard deviation from the mean.

This is one "tail" of the distribution.

MODERATE MEAN AND STANDARD DEVIATION

MODERATE MEAN AND SMALL STANDARD DEVIATION

ANY SIZED MEAN AND A LARGE STANDARD DEVIATION

PRACTICE PROBLEMS

11-11. Draw two graphs of the data from Problem 11-4 one as a histogram and one as a line graph connecting the points: 2, 4, 6, 7, 8, 9, 10, 11, 12, 13.

11-12. A large collection of talons of fossil raptors were subjected to measurements, of which some length data are shown here. One grouping of the categories waa done at every 1 cm, and the other every 3 cm. Draw the graphs

Cm	0.5	1.5	2.5	3.5	4.5	5.5	6.5	7.5	8.5	9.5
Number	1.5	5	5	10	5	8	11	9	4	2
V X N	0.8	7.5	12.5	0.35	22.5	44	71.5	67.5	34	19

Cme	1	3	5	7	9	0
Number	15	13	13	20	6	0
V X N	15	45	65	140	9	0

The Polygene Hypothesis

> Quantitative traits are characterized by descriptive statistics such as the mean and the standard deviation. They comprise a significant number of the traits encountered in nature, and in agriculture so they require an explanation of the mechanism of inheritance.

The **polygene hypothesis** for the inheritance of quantitative characters is based on the results of a particular cross. This or similar crosses have been done with beans, wheat grains, sunflower seeds, tobacco flowers and many aspects of corn. The polygene hypothesis has also been studied in *Drosophila*, guinea pigs and people. Here is a summary of the hypothesis:

It became clear that at least some quantitative traits were inherited in a pattern like that which Mendel observed, but with these important differences:

1. Mendel used a plant with sharply differing traits, whereas the quantitative geneticists work with traits that vary about the mean so as to fall within a normal curve.
2. Both the mean and the standard deviation are aspects of a phenotype that results from an interaction between the genotype and the environment.
3. The genes involved in such inheritance are said to have additive effects – they each add to the quantitative phenotype of interest. In addition, the environment has a large effect compared to these genes.
4. It is sometimes possible to figure out how many loci are involved in the formation of a quantitative trait if an F_1 and F_2 cross can be made. If the sample size is large enough, the proportion of the least frequent phenotypic class in the F_2 will tell us how many loci we are looking at.

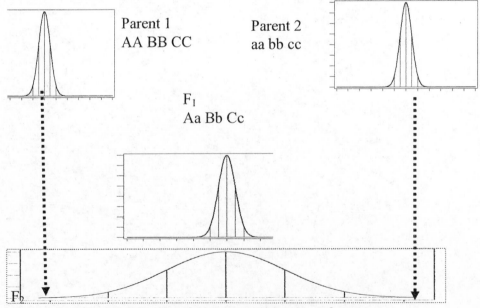

Wide distribution due to recombination of genes.
The extremes of F_2 overlap the parental distributions.

A simple explanation for these crosses is to assume that the variability we see in both the means and standard deviations is due to the independent assortment of alleles of one or more loci. Suppose that every uppercase allele of one of these loci will increase the length of the ear by 0.5 cm beyond a basal length of 5 cm. Thus, a genotype of AA Bb cc would be basal of 5 cm + 1.5 cm = 6.5 cm due to having three uppercase alleles.

The cross is a simple one: A cross is preformed of two true breeding (homozygous) flies to get heterozygotes for each locus of interest. Then a cross is performed of heterozygotes to get all the possible F_2 genotypes.

Note these aspects of the cross:

1. The variability in the F_1 is about the same magnitude as the parents.
2. But the variability in the F_2 is much greater.
3. A normal curve symbolizes the trait being studied, not a single, clearly defined characteristic, but a character with a central tendency and variation. The position of this symbol on the page is an indication of the mean of each participant in the cross.
4. The mean of the F_1 is intermediate to the parentals.
5. The smallest categories among the F_2 are AA BB CC and aa bb cc, each 1/64 of the total F_2 progeny.

The result of this cross on a system of three loci each with two alleles is shown by an 8 X 8 Punnett square. The number on the second row of each box is the number of uppercase alleles in that zygote. With these numbers, the weight of each class of beans can be calculated.

	ABC	ABc	AbC	Abc	aBC	aBc	abC	Abc
ABC	AA BB CC 6	AABBCc 5	AABbCC 5	AABbCc 4	AaBBCC 5	AaBBCc 4	AaBbCC 4	AaBbCc 3
ABc	AABBCc 5	AABBcc 4	AABbCc 4	AABbcc 3	AaBBCc 4	AaBBcc3	AaBbCc 3	AaBbcc 2
AbC	AABbCC 5	AABbCc 4	AAbbCC 4	AAbbCc 3	AaBbCC 4	AaBbCc 3	AabbCC 3	AabbCc 2
Abc	AABbCc 4	AABbcc 3	AAbbCc 3	AAbbcc 2	AaBbCc 3	AaBbcc 2	AabbCc 2	Aabbcc 1
aBC	AaBBCC 5	AaBBCc 4	AaBbCC 4	AaBbCc 3	aaBBCC 4	aaBBCc 3	aaBbCC 3	aabbCc 1
aBc	AaBBCc 4	AaBBcc 3	AaBbCc 3	AaBbcc 2	aaBbCc 2	aaBBcc 2	aaBbCc 2	aaBbcc 1
abC	AaBbCC 4	AaBbCc 3	AabbCC 3	AabbCc 2	aaBbCC 3	aaBbCc 2	aabbCC 2	aabbCc 1
Abc	AaBbCc 3	AABbcc 3	AabbCc 2	Aabbcc 1	aaBbCc 2	aaBbcc 1	aabbCa 1	aabbcc 0

Solved Problem 11-13. How does counting parental genotypes tell how many loci are involved in a quantitative trait?

Answer. If there had been two loci instead of three, there would have been fewer genotypes. As in Chapter 1, the number of gametes produced by an organism is 2^n, where n = number of heterozygous loci. For a two locus system, the number of gametes is $2^2 = 4$. If two such organisms were mated, the results are informative. The most infrequent classes of offspring have the same genotypes as the parental types.

Number of Loci in Heterozygotes	Number of Possible Gametes $[\,2^n\,]$	Number of Cells in the Punnett Square $[2^n]^2$	Fraction of Offspring Like the Original Parents
1	2	4	1/4 AA, 1/4 aa
2	4	16	1/16 AABB 1/16 aa bb
3	8	64	1/64 AABBCC 1/64 aa bb cc
4	16	256	1/256 AABBCCDD 1/256 aa bb cc dd
5	32	1024	1/1024 AABBCCDDEE 1/1024 aa bb cc dd ee

PRACTICE PROBLEMS

11-14. Suppose you had a mammal with the genotype AaBb. In this case, let b designate a recessive lethal allele that kills the embryo if bb while they are still eggs, so we won't see them among the progeny. Cross pairs of Aa Bb. What will be the phenotypes of the next generation?

Assume that the effects of A and B are additive for weight, each uppercase allele adding 1 ounce the basal weight of 16 ounces. In the Punnett square, the first line of a cell is the genotype, the second line tells whether the lethal gene is expressed or not, and the third line is the weight calculated on the basis that each uppercase allele contributes.

CHAPTER 12

A COLLECTION OF REVIEW PROBLEMS

Throughout this guide, you have encountered problems within a specific context, which gave you some idea of how to go about solving the problem. Ordinarily, a geneticist would not have this kind of context. So in that spirit, here is a "bonus round". In this chapter, you will find a variety of problems, covering all of the approaches you have learned if you have applied yourself to this book. Use this set of practice problems as a review, and to see how much you've actually learned.

R-1 through R-5 give data from several different genetics experiments involving *Drosophila*. Answer each of the following questions where possible.

Number of loci?
Number of alleles?
Dominant and recessive?
Autosomal or sex-linked?
Independent assortment or linkage?
If linkage, what are the map order and map distances.

Give the genotypes of the parents and progeny whenever possible.

Don't forget these facts about *Drosophila:* crossing over only occurs in females, and females are XX while males are XY.

The answers are in the answer section, starting on page <>.

R-1. When two *Drosophila* with wild type (brick red) eyes were crossed, the **male** offspring included 187 with raspberry-colored eyes and 194 with wild type eyes. All 400 **females** were wild type.

R-2. *Drosophila* with scarlet eyes (redder than wild type) were crossed to *Drosophila* with brown eyes (browner than wild type). The F_1 all had wild type eyes. The F_2 included 440 with wild type eyes, 137 with scarlet eyes, 150 with brown eyes, and 48 with white eyes.

R-3. The dominant allele for Bar eye in *Drosophila* reduces the number of facets in the compound eye. The recessive allele for carnation results in eyes with a redder than normal color. Both are X-linked genes. A female fly heterozygous for both genes was crossed to a male with eyes of carnation color but normal shape. The male progeny included 470 with both the Bar and carnation phenotypes, 465 with normally shaped and colored eyes, 20 with Bar eyes of normal color, and 28 with normally-shaped carnation eyes. Female progeny included 500 with Bar and carnation eyes, 475 with eyes normally shaped and colored, 18 Bar eyes, but with normal color and 25 with normally shaped carnation eyes.

145

R-4. The Eyeless allele reduces the number of facets in the compound eye so that only a few are left. Purple changes the eye color (to purple!). All the offspring of a cross of two wild type flies were wild type. Individual offspring were then test crossed. Four kinds of results occurred in about equal amounts.

A 1:1:1:1 ratio of wild type, eyeless and purple, eyeless with normal color, and purple of normal shape.

All wild type

A 1:1 ratio of wild type to purple

A 1:1 ratio of eyeless to wild type.

R-5. An allele at the white locus (w) results in no eye color pigment being produced. Another allele w reduces the pigment to 10 % of that in wild type. Heterozygous flies for w and w have a pigment level halfway between the two alleles. Wild type is completely dominant to both w and w. A female heterozygous for w and w is crossed to a wild type male. All the female progeny have wild type eyes, while half the male progeny are white and half are apricot.

Questions R-6 through R-10 are about genetics in humans and human families. Reason like a geneticist to answer the questions.

R-6. The affliction called retinitis pigmentosa has two genetic causes. It is sometimes due to a recessive allele a (when homozygous, of course) or by a dominant R at an independently assorting locus. To avoid the disease, one must have at least one dominant A and also be homozygous for r.

Two normal parents have a son with reitinits pigmentosa, and this son marries a woman who is a double heterozygote.

A) What is the woman's phenotype?

B) What proportion of their offspring will be free of the disease?

R-7. A) If the only child in a family has type A blood, and the mother has O type, what genotypes are possible for the father?
B) What if there are four children with one with type O blood, one with type A, one with type B, and one with type AB and the mother had type A? What genotypes could the father be then?

R-8. The pedigree that follows this problem is for a rare recessive condition. Obviously it must have been carried in heterozygous condition through several generations. If person number I-1 was a heterozygote, what other people in the pedigree

<u>must</u> have been heterozygotes also? (For example, I-2 <u>could</u> be heterozygous, but would not necessarily have to be since I-1 could have passed her copy to the next generation. Just identify the ones that absolutely <u>must</u> be heterozygotes.)

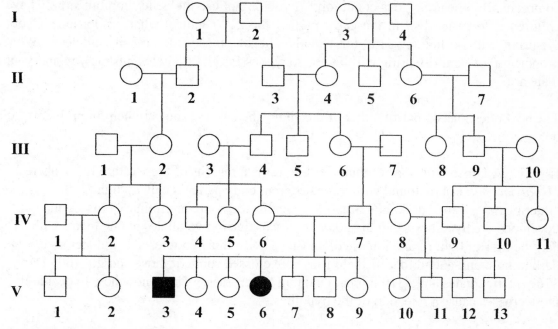

R-9. Huntington's chorea is a neurological affliction that produces its first symptoms in people in their middle years. It causes degeneration of motor control regions of the brain, resulting in progressive debilitation over a 15 to ultimately 25 year period. The disease is fatal. It has been found that if a child's mother or father develops Huntington's chorea, the child has a 50% change of developing it also.

A) What is the simplest genetic explanation for the cause of the disease?

B) In neither of a person's parents have Huntington's chorea but one grandparent did, what is the probability that this person will develop the disease?

C) Suppose a woman had a brother with Huntington's chorea. The children's' parents had died in an auto accident when the children were very young; therefore it was not known whether one of both of the parents would have developed Huntington's chorea in middle age. What is the minimum number of parents that could have had the disease?

R-10. Suppose in the future you can go to the gene shop and find out which genes you have and how many of them are homozygous and heterozygous. You just got back from the gene shop where you found that you and your mate have 20 recessive alleles in common, all in heterozygous condition. If you do not use the gene shop but simply have children as they used to in the early part of the 21st century, what proportion of your offspring will be homozygous for exactly one of those 20 recessive alleles? What proportion of your offspring will be heterozygous for all 20 alleles, just as you and your mate are?

The next questions do not fit either of the above themes, but they are helpful problems to do.

R – 11. A Drosophila species that lives in tropical regions of the western hemisphere, *D. polymorpha* can be found with three different body patterns called "light," "intermediate," and "dark." A Brazilian scientist did some preliminary crosses and developed the hypothesis that the body color was due to two alleles at one locus, with dark being EE, light ee and intermediate Bb. To test this hypothesis he did a cross of dark to light and got all intermediate flies. He crossed some of those and got 1605 Dark, 3767 intermediates and 1310 lights. Diagram the crosses. Then show how to decide if these crosses cause a rejection of the hypothesis.

R-12. Refer to problem R-11. The geneticists studying this species counted 8070 flies. Of these, 3969 were dark, 3174 were intermediate and 927 were light. Is this population in Hardy-Weinberg equilibrium?

BIBLIOGRAPHY

For this edition of *A Problem-Based Guide to Basic Genetics,* new practice problems based on research papers in the scientific literature have been added. The problems are indicated in the chapters by an asterisk before the problem number. Often the results in the research paper had to be simplified somewhat to make a good problem, but the references are included here in case you would like to see the data that produced the problem. [The material in square brackets indicates which problems are based on that paper.]

Astolfi, P., M. Cuccia, and M. Martinetti. 2001. Paternal HLA genotype and offspring sex ratio. Human Biology 73 (2): 315-319. [3-17, pg. 38]

Bed'Hom, Bertrand, Phillippe Coullin, Zuzana Guillier-Gencik, Sibyle Moulin, Alain Bernheim, and Vitaly Volobouev. 2003. Characterization of the atypical karyotype of the black-winged kite Elanus caeruleus (Falconiformes: Accipitridae) by means of classical and molecular cytogenetic techniques. Chromosome Research 11: 335-343. [8-3, pg. 81]

Benedict, M.Q., L.M. McNitt, and F.H. Collins. 2003. Genetic traits of the mosquito, Anopheles gambiae: red stripe, frizzled, and homochromy 1. Journal of Heredity 94(3): 227-235. [1-15, pg. 6, 7-19, pg. 78]

Bosland, P.W. 2002. Inheritance of a novel flaccid mutant in Capsicum annuum. Journal of Heredity 93(5): 380-382. [1-20, pg. 7, 1-45, pg. 17, 1-50, pg. 19]

Comai, Luca, Anand P. Tyagi, and Martin A. Lysak. 2003. FISH analysis of meiosis in Arabidopsis allopolyploids. Chromosome Research 11: 217-226. [8-14, pg. 89]

Eberle, Michael A., Roland Pfizer, Kay L. Pogue-Geile, Mary P. Bronner, David Crispin, Michael B. Kinney, Richard H. Duerr, Leonid Kruglyac, David C. Whitcom, and Teresa A. Brentnall. 2002. A new susceptibility locus for austosomal dominant pancreatic cancer maps to chromosome 4q32-34. American Journal of Human Genetics 70: 1044-1048. [5-22 C), pg. 60]

Favor, Jack, Heiko Peters, Thomas Hermann, Wolfgang Schmahl, Bimal Chatterjee, Angelica Neuhänuser-Klaus, and Rodica Sandulache, 2001. Molecular Characterization of Pax62ITet1 through Pax6'°'': An extension of the Pax6 allelic series and the identification of two possible hypomorph alleles in the mouse Mus musculus. Genetics 159: 1689-1700. [2-12, pg. 28, 3-6, pg. 33, 7-10, pg. 73]

Kaiser, Hinrich, Claus Steinlein, Wofgang Feichtinger, and Michael Schmid. 2003. Chromosome Banding of Six Dendrobatid Frogs (Colostethus, Mannophryne). Herpetologica 59 (2): 203-2 18. [8-7, pg. 83]

Kern, A. J., T. M. Myers, M. Jasieniuk, B.G. Murray, B.D. Maxwell, and W.E. Dyer. 2002. Two recessive gene inheritance for triallate resistance in Avena fatua L. Journal of Heredity 93(1): 48-50. [3-25, pg. 42]

Kijas, J.W., B.J. Miller, S.E. Pearce-Kelling, G.D. Aguirre, and G.M. Acland. 2003. Canine models of ocular disease: outcross breedings define a dominant disorder present in the English mastiff and bull mastiff dog breeds. Journal of Heredity 94(1): 27-30. [1-5, pg. 2, 5-15, pg. 56]

Matsubara, Kazumi, Chizuko Nishida-Umehara, Asato Kuroiwa, Kimiyuki Tsuchiya, and Yoichi Mastuda. 2003. Identification of chromosome rearrangements between the laboratory mouse (Mus musculus) and the Indian spiny mouse (Mus platythrix) by comparative FISH analysis. Chromosome Research 11: 5 7-64. [8-9, pg. 85]

Max, Marianna, Y. Gobi Shanker, Liquan Huang, Minqing Rong, Zhan Liu, Fabien Compagne, Hare! Weinstein, Sami Damak, and Robert F. Margolskee. 2001. Taslr3, encoding a new candidate taste receptor, is allelic to the sweet responsiveness locus Sac. Nature Genetics 28(5): 58-63. [1-27, pg. 9]

Pedrosa, Andrea, Niels Sandal, Jens Stougaard, Dieter Schweitzer, and Andreas Bachmair. 2002. Chromosomal map of the model legume Lotus japonicus. Genetics 161: 1661-1672. [6-15, pg. 66, 6-16, pg. 66, 6-21, pg. 69, 7-5, pg. 71]

Piao, Xianhua, Lina Basel-Vanagaite, Rachel Straussberg, P. Ellen Grant, Elizabeth W. Pugh, Kim Doheny, Beth Doan, Susan E. Hong, Yin Yao Shugart, Christopher A. Walsh. 2002. An autosomal recessive form of bilateral frontoparietal polymicrogyria maps to chromosome 16q12.2-21. American Journal of Human Genetics 70: 1028- 1033. [5-22 B) pg. 60]

Rankinen, Tuomo, Louis Pérusse, and Rainer Rauramaa. 2002. The human gene map for performance and health-related fitness phenotypes: the 2001 update. Medicine and Science in Sports and Exercise 34(8): 12 19-1233. [6-6, pg. 62, 6-12, pg. 64]

Schlipalius, David I., Qiang Cheng, Paul E. B. Reilly, Patrick J. Collins, and Paul R. Ebert. (2002). Genetic linkage analysis of the lesser grain borer Rhyzopertha dominica identifies two loci that confer high-level resistance to the fumigant phosphine. Genetics 161: 773-782. [2-6, pg. 25]

Schwarz-Sommer, Zsuzsanna, Eugenia de Andracle Silva, Rita Berndtgen, WolfEkkehard Lonnig, Andreas Muller, Ingo Nindl, Kurt Stüber, Jorg Wunder, Heinz Saedler, Thomas Gübitz, Amanda Borking, John F. Golz, Enrique Ritter, and Andrew Hudson. (2003). A linkage map of an F2 hybrid population of Antirrhinum majus and A. molle. Genetics 163: 699-710. [6-21, Pg. 69)
108

Srevalli, Y., R. N. Kulkarni, and K. Baskaran. 2002. Inheritance of flower color in periwinkle: orange-red corolla and white eye. Journal of Heredity 93(1): 55-58. [1- 1l,pg.4, I-32,pg. 11, 1-39, pg. 15]

Tartaglia, Marco, Kamini Kolidas, Adam Shaw, Xiaoling Song, Dan Musat, Ineke van der Burgt, Han G. Brunner, Deborah R. Bertola, Andrew Crosby, Andra Ion, Raju S. Kucherlapati, Steve Jeffery, Michael A. Patton, and Bruce D. Gelb. 2002. PTPN2 mutations in Noonan syndrome: molecular spectrum, genotype-phenotype correlation, and phenotypic heterogeneity. American Journal of Human Genetics 70: 1555-1573. [5-16,pg. 56]

Turley, R. B. and R.H. Kloth. 2002. Identification of a third fuzzless locus in upland cotton (Gossypium hirsutum L.) Journal of Heredity 93(5): 359-364. [1-56, pg. 21, 3-24, pg. 42]

Villard, Laurent, Kaline Nguyen, Carlos Cardoso, Christa Lev Martin, Ann M. Weiss, Mara Siffry-Platt, Arthur W. Grix, John M. Graham, Jr., Robyn M. Winter, Richard L. Leventer and William B. Dobyns. 2002. A locus for bilateral perisylvian polymicrogyria maps to Xq28. American Journal of Human Genetics 70(4): 1003- 1008. [5-22 A), pg. 59]

Wang, P. Jeremy, John R. McCarrey, Fang Yang, and David C. Page. 2001. An abundance of X-linked genes expressed in spermatogonia. Nature New Genetics 27: 422-426. [4-2, pg. 44, 4-8, pg. 46.4-14, pg. 49] 109

ANSWERS TO PRACTICE PROBLEMS

CHAPTER 1 - SIX SYSTEMATIC STEPS

1-2. P: brown X white

F$_1$: all brown

F$_2$: 1 white, the others brown

1-3. P: yellow, plain X white, *marked*

F$_1$: yellow, marked   yellow, plain   white, marked   white, plain
      65   56   61   59

1-4. P: golden-eyed   X   green-eyed

F$_1$:   some golden-eyed; some green-eyed

*1-5.   First cross

P: PRA male X Normal female

F$_1$:   PRA males PRA females Normal males Normal females
          0   1   2   1

Second cross

P: Normal male X PRA female

F$_1$:   PRA males PRA females Normal males Normal females
          1   1   3   1

1-8. If the Huntington's locus is *H* for disease and *h* for normal and lack of cystic fibrosis is *C* and the presence of cystic fibrosis is *c*, then

A) *Hh CC* (or rarely *HH CC*)   B) *hh CC*   C) *hh cc*

1-9. Let *E* = wild type straw-colored body and *e* = ebony body color

P: *EE* X *ee*

F$_1$:   all *Ee*

    152

1-10. The letter G seems natural for this cross. Let G be the dominant allele and g the recessive. The golden-eyed lacewings are Gg. The green-eyed lacewings are gg.

*1-11. A) $Rr\ Ww\ Ee$ B) $RR\ Ww\ ee$ C) $rr\ ww\ ee$

1-13. Black sheep and spotted goats were true-breeding, while solid-colored goats and white sheep were not.

1-14. The parents are heterozygous. They gave birth to pups unlike themselves. The pups, however, are homozygous (true-breeding). When two pups mate with each other, they produce silver-backed pups.

*1-15. *Homochromy I* must be due to a recessive. Pure lines for dominant mutations are harder to find because some individuals expressing the allele will be heterozygous. If they always form pure lines in one generation, they must all be hh to start with.

1-18. Yes, we can know the genotypes of the true-breeding strains since "true-breeding" means "homozygous." If we let the symbol B stand for the allele for white sheep and b stand for the allele for black sheep, then Jake's black sheep were bb and the white sheep were $B\_$ (where the blank means that the other allele could be either B or b). Likewise, if S is the allele for solid colors and s is the allele for spotted, Jake's spotted goats are ss, and the solid goats are $S\_$.

1-19. If peroneal muscular atrophy were caused by a recessive allele, it would be possible for two parents without the disease to give birth to an affected child if both of them were heterozygous. If at least one parent has to have the disease for the child to inherit it, then the allele must be dominant.

*1-20. Here are the three crosses described in the problem:

flaccid X flaccid	KRG X KRG	KRG X flaccid
All flaccid	All normal	All normal

Flaccid offspring only come from two flaccid parents. Flaccid must be recessive (symbolize f) and normal dominant (F). Thus, we know these genotypes for certain. KRG crossed to KRG always gives normal, so KRG strain is probably homozygous too.

flaccid X flaccid	KRG X KRG	KRG X flaccid
ff ff	$F\_$ $F\_$	$F\_$ ff
	(FF) *(FF)*	*(FF)*
All flaccid	All normal	All normal
Ff	FF	Ff

1-23. Since the flies have the dominant phenotype, they must either be *WW* or *Ww*. We can know for certain that they have at least one *W* allele. So we could symbolize their genotype as *W_*, with the blank standing for either *W* or *w*.

1-24. The runt litter mates are *rr*, which means that each parent was *Rr*. Since Honey is not a runt, he got an *R* allele from one parent, but he could have received either an *R* or an *r* from the other parent. So he is either *RR* or *Rr*.

1-25. Cross Honey to a runt pig (*rr*). If Honey is *RR*, all the offspring will get an *R* from Honey and will therefore all be normal. But if Honey is *Rr*, some offspring will get an *r* from Honey and an *r* from the runt, so some of Honey's offspring will be *rr* runts.

1-26. Genotypes A, D, and E are homozygous for at least one of the recessive alleles, so they are each sensitive to potassium. Genotypes B and C have at least one dominant allele at each locus, so they are resistant.

*1-27. Here are the possible crosses:

(1)	(2)	(3)
TT X *TT*	*tt* X *tt*	*TT* X *tt*
All *TT*	All *tt*	All *Tt*

T = taster All taster *T* = taster All non-taster *T* = taster All taster
T = non-taster All non-taster *T* = non-taster, all taster *T* = non-taster All non-taster

(4)	(5)	(6)
Tt X *TT*	*Tt* X *tt*	*Tt* X *Tt*
1/2 *TT* and 1/2 *Tt*	1/2 *Tt* and 1/2 *tt*	1/4 *TT* 1/2 *Tt* 1/4 *tt*

T = taster All taster T = taster 1/2 of each T = taster 3/4 tasters
T = non-taster All non-taster T = non-taster 1/2 of each T = non-taster 3/4 non-tasters

Crosses 1 and 2 are not very useful. Neither is Cross 5, since it gives the same results for each possibility. Cross 3 is direct. If you know the parents are each homozygous, then Cross 3 tells you which allele is dominant. Cross 6 can be helpful if you have enough offspring. Then the smaller phenotype detected in 1/4 of the cases is the recessive phenotype.

1-29. All are true-breeding (homozygous), so if their genotype is *WW*, all the gametes will contain allele *W*. If their genotype is *ww*, all their gametes will contain *w*.

1-30. Both parents are heterozygous, so each will produce gametes with *A* and *a* alleles in equal proportions.

1-31. In heterozygous males, the allele will be lost. The gametes are haploid. If that allele is lethal, as in this case, it will kill the sperm. No sperm with that lethal allele will be able to fertilize any eggs. A heterozygous female could produce eggs with the gene that would then be passed on to the next generation. But it could only be maintained in heterozygous females because male parents will never pass the gene on. So if there is a way to identify the females with the allele, the gene could be maintained.

*1-32. *EE* will produce *E* gametes.
ee will produce *e* gametes.
Ee will produce *E* and *e* gametes.

1-36. Both parents must be heterozygous for cleft palate. Let *P* be the normal allele and *p* be the cleft palate allele. Each hog can produce both *P* and *p* gametes in a 1:1 ratio.

1-37. There are four heterozygous loci, so there will be 2^4, or 16, possible gamete combinations:

EFTV	*eFTV*
EFTv	*eFTv*
EFtV	*eFtV*
EFtv	*eFtv*
EfTV	*efTV*
EfTv	*efTv*
EftV	*eftV*
Eftv	*eftv*

1-38. If there are 10 heterozygous loci, there will be 2^{10} or 1024 possible allele combinations in gametes. Aren't you glad I didn't ask you to figure out what they all are?

*1-39. A) Parents are each homozygous for all loci, so they can each produce only one type of gamete: One parent will produce *e R w 0* and the other *E r W o*. The F$_1$ will be heterozygous at all four loci, (*Ee Rr Ww Oo*) so it will produce $2^4 = 16$ gamete types. To obtain all the gametes with the alternating list method for the F$_1$, first make a list of eight *E* and another list of eight *e*. I made my list sideways to take up less room:

E   E   E   E   E   E   E   E
e   e   e   e   e   e   e   e

Then put a list of four *R* and four *r* next to each list of *E* or *e:*

ER   ER   ER   ER   Er   Er   Er   Er
eR   eR   eR   eR   er   er   er   er

Then put a list of two *W* and two *w* as shown below and then finally alternating *O* and *o* lists as shown:

ERW   ERW   ERw   ERw   ErW   ErW   Erw   Erw

$e\,R\,W\quad e\,R\,W\quad e\,R\,w\quad e\,R\,w\quad e\,r\,W\quad e\,r\,W\quad\quad e\,r\,w\quad\quad e\,R\,w$

And finally, the *O* and *o* alleles:

$E\,R\,W\,O\quad E\,R\,W\,o\quad E\,R\,w\,O\quad E\,R\,w\,o\quad E\,r\,W\,O\quad E\,r\,W\,o\quad E\,r\,w\,O\quad E\,r\,w\,o$
$E\,R\,W\,O\quad e\,R\,W\,o\quad e\,R\,w\,O\quad e\,R\,w\,o\quad e\,r\,W\,O\quad e\,r\,W\,o\quad e\,R\,w\,O\quad e\,r\,w\,o$

Progeny A, B, C and D work out this way with the branch method:

B) *Ee RR Ww*

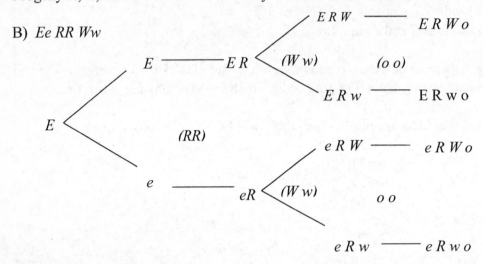

C) This is homozygous for each locus. Only one type of gamete will be produced, ERWo.

D) ee rr Ww oo

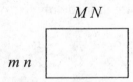

1-42. Each individual only produces one kind of gamete, so the Punnett square is simple:

$M\,N$

$m\,n$

1-43. *Bb Ff Gg* individuals will produce eight allele combinations in gametes. So there will be eight gametes on each side of the square. See diagram on next page.

	BFG	BFg	BfG	Bfg	bFG	bFg	bfG	bfg
BFG								
BFg								
BfG								
Bfg								
bFG								
bFg								
bfG								
Bfg								

1-44. The one parent with 14 heterozygous loci would produce 2^{14}, or 16,384 types of gamete and the other parent with three heterozygous loci would produce eight kinds of gamete. Since a Punnett square has gametes along each side, this one would be 16,384 squares in one direction and eight squares in the other.

*1-45. There are four crosses described, so we'll set up a square for each. Let f = the f

F F X f f   Selfed: F f X F f   F f X f f   F f X F F

	F
F	Ff

	F	f
F	F F	F f

	F	f
F	Ff	Ff

	F	F
F	F F	Ff

1-47.

1-48.

	M N
mn	MmNn

	BFG	BFg	BfG	Bfg	bFG	bFg	bfG	bfg
BFG	BB FF GG	BB FF Gg	BB Ff GG	BB Ff Gg	Bb FF GG	Bb FF Gg	Bb Ff GG	Bb Ff Gg
BFg	BB FF Gg	BB FF gg	BB Ff Gg	BB Ff gg	Bb FF Gg	Bb FF gg	Bb Ff Gg	Bb Ff gg
BfG	BB Ff GG	BB Ff Gg	BB ff GG	BB ff Gg	Bb Ff GG	Bb Ff Gg	Bb ff GG	Bb ff Gg
Bfg	BB Ff Gg	BB Ff gg	BB ff Gg	BB ff gg	Bb Ff Gg	Bb Ff gg	Bb ff Gg	Bb ff gg
bFG	Bb FF GG	Bb FF Gg	Bb Ff GG	Bb Ff Gg	bb FF GG	bb FF Gg	bb Ff GG	bb Ff Gg
bFg	Bb FF Gg	Bb FF gg	Bb Ff Gg	Bb Ff gg	bb FF Gg	bb FF gg	bb Ff Gg	Bb Ff gg
bfG	Bb Ff GG	Bb Ff Gg	Bb ff GG	Bb ff Gg	bb Ff GG	bb Ff Gg	bb ff GG	Bb ff Gg
bfg	Bb Ff Gg	Bb Ff gg	Bb ff Gg	Bb ff gg	bb Ff Gg	bb Ff gg	bb ff Gg	Bb ff gg

1-49. The square is 16,384 X 8. So the number of intersections is 131,072.

*1-50. Here are the Punnett squares filled out with the phenotypes identified:

F F X f f   Selfed: F f X F f   F f X f f

F f X F F

	F
F	Ff Normal

	F	f
F	FF Normal	Ff Normal
F	F f Normal	f f Flaccid

	F	f
f	Ff Normal	ff Flaccid

	F	f
F	FF Normal	Ff Normal

CHAPTER 2 - GOING BEYOND THE SIX STEPS

2-3. In the cross of *Ff Gg Hh* X *ff gg hh*, the triple heterozygote will produce eight different gamete types, and the triple homozygote will produce only one kind (*f g h*). So each of the possible offspring will receive one of the eight possible gametes from the heterozygote and one *f g h* allele combination in the gametes from the homozygous recessive parent. All of the possible offspring are listed below.

Gametes from Ff Gg Hh		gamete from ff gg hh	Possible offspring	
F G H	*f G h*	*f g h*	*Ff Gg Hh*	*ff Gg Hh*
F G h	*f G h*		*Ff Gg hh*	*ff Gg hh*
F g H	*f g H*		*Ff gg Hh*	*ff gg Hh*
F g h	*f g h*		*Ff gg hh*	*ff gg hh*

2-4. Let *D* be the allele for polydactyly and *d* the recessive allele. Let *p* be the allele for phenylketonuria and *P* the allele for lack of phenylketonuria. Since the man's father had phenylketonuria, the father must have been *pp*, so the man is *Pp*. The man doesn't have polydactyly either, so he is *dd*. The woman has polydactyly, so she is *Dd*. Since the problem specifies that she has no phenylketonuria allele, she is also *PP*. Here is a diagram of what we know about this cross:

P: *Pp dd*   X   *PP Dd*

Gametes: 1/2 *Pd* 1/2 *pd*   1/2 *PD* 1/2 *Pd*

Punnett square:

	Pd	*pd*
PD	*PP Dd* polydactyly	*PpDd* polydactyly
Pd	*PP dd* Neither condition	*Pp dd* neither condition

There is no chance that any offspring will have both conditions, nor is there a chance that any offspring will have phenylketonuria. Half the offspring can be expected to have polydactyly.

2-5. The investigators have the two single mutant lines *LL pp* and *ll PP*. Cross each of them to every unknown. Each type of unknown will give different results. Diagram the crosses to convince yourself of the results summarized in the table. Two crosses are diagrammed to give you an example of what to do.

Examples:

P: *ll PP* (known)   X   unknown (if *ll PP*)
Gametes: all *lP*   all *lP*
F₁: all *ll PP* light-eared

P: *ll PP* (known)   X   unknown (if *LL pp*)
Gametes: all *lP*   all *Lp*
F₁: all *Ll Pp* normally eared

Known		Unknowns	
Testers	*If ll PP*	if *LL pp*	if *ll pp*
Ll PP	all light ears	no light ears	all light ears
LL pp	no light ears	all light ears	all light ears

So the double mutants are the unknowns that have light-eared progeny in both crosses to the testers. The *ll* single mutant is the one that produces light-eared progeny when crossed to known *ll* testers, and the *pp* single mutant is the one that produces light-eared progeny when crossed to the known *pp* tester.

*2-6. Cross

A) *P6P6* X *p6p6*   F₁   P6p6   F₂   1 P6P6 : 2 P6p6 : 1 p6p6
                              No resistance   None   None   12.5 fold

B) P5P5 X p5p5   F₁   P5p5   F₂ 1 P5P5 : 2 P5p5 : 1 p5p5
                          No resistance   None   None   50 fold

C) P5P5 p6p6 · X p5p5 P6P6
                    F₁ P5p5 P6p6   No resistance
                    F₂   9 P5 __ P6 __   No resistance
                            3 p5p5 P6 __   50-fold
                            3 P5 __ p6 p6   12.5-fold
                            1 p5 p5 p6 p6   250-fold

2-9. The ratio is nearly 9:3:3:1, which is characteristic of a cross of two parents heterozygous for two pairs of independently assorting alleles. So the parents were

$$Dd\,Nn \qquad\qquad X \qquad\qquad Dd\,Nn$$

2-10. Write what we know of the genotypes of Izzy and Lucinda.

Izzy   Lucinda

*B_ ww F_*   *bb W_ ff*

If Lucinda's other parent had big white spots, weak hooves and at least one *f* allele (*B_ W_ _f*), then it is possible that Izzy could be Lucinda's parent.

2-11. Look for familiar ratios. The F₂ progeny approximate a 9:3:3:1 ratio, characteristic of a cross of two double heterozygotes with each locus having a dominant and recessive allele. So let the F₁ be *Aa Bb*. Crossing two of those gives the F₂

            311   *A_ B_* (normal green)
            107   *aa B_* (lighter green)
            123   *A_ bb* (pale green)
             32   *aa bb* (yellow green)

The 1/16 class and the 9/16 class are unambiguously *aa bb* and *A_ B_*. Lighter green is homozygous for one recessive allele, and pale green for the other. If we call lighter green *aa B_*, then pale green must be *A_ bb*. This is entirely arbitrary, however; an *A_ bb* could be lighter green and *aa B_* could be pale green.

Strain 12 must be the true-breeding version of *e* genotype of the pale green genotype (*AA bb*), and strain 15 must be *aa BB*.

CHAPTER 3 - PROBABILITY AND STATISTICS

3-6. If the parents were heterozygous at one locus, the probability of giving rise to homozygous recessive offspring is 1/4. For 10 independently assorting loci, the probability of homozygous loci is $(1/4)^{10}$ = 0.000000915 or approximately 0.000001.

3-7. We learn that red and white (*r*) is recessive to black and white (*R*). The presence of horns is due to a recessive allele (*h*), and the hornless condition is dominant (*H*). Then diagram the cross:

Black and white , horned   X   red and white, hornless
        Rr   hh   rr   Hh

The farmer wants red and white and hornless like the father.
Take each locus separately:

           Rr X rr   hh X Hh
        1/2 Rr and 1/2 rr   1/2 Hh and 1/2 hh

The desired bulls are rr (1/2) Hh (1/2) so the probability is 1/2 X 1/2 = 1/4.

The farmer is going to do 15 of these matings to produce 30 calves. Of those, 1/4 of 30 = 7.5, should be about what he would get.

3-8. Consider crosses of all eight loci separately:

Aa X *Aa* → 1/4 *AA*, 1/2 *Aa* and 1/4 *aa*   *Ee* X *Ee* → 1/4 *EE*, 1/2 *Ee*, 1/4 i
BB X *BB* → all *BB* (probability is 1.0)   *ff* X *ff*→ all *ff* (probability of 1.0)
cc X *cc* → all *cc* (probability is 1.0   *gg* X *gg* → all *gg* (probability of 1.0)
Dd X *Dd* → 1/4 *DD*, 1/2 *Dd*, 1/4 *dd*   *HH* X *HH* → all *HH* (probability of 1.0)

To find the probability of these eight independent crosses, multiply their probabilities.
For *AA BB cc Dd Ee ff gg HH*   1/2 X 1 X 1 X 1/2 X 1/2 X 1 X 1 X 1 = 1/8.

*3-9. P: N2 n2 N3 n3   X   N2n2 N3n3

Fuzzless progeny include *n2n2 n3n3*. Progeny with any *N* allele, whether *N2* or *N3* will have fuzz. The probability of a parent producing an *n3* gamete is 1/2. The probability of an *n3n3* offspring is (1/2) (1/2) = 1/4. By similar reasoning, the probability of an *n3 n3* is also 1/4. So the probability that an *n2n2 n3n3* will be produced is (1/4) X (1/4) = 1/16.

3-12. A man has sickle cell anemia (*ss*) and is heterozygous for cystic fibrosis (*Cc*). The woman is *Ss Cc*. A cross is *ss Cc* X *Ss Cc*. Approach one locus at a time.

First, they want to know about girl children, so about 1/2 males and 1/2 females.

The sickle cell anemia cross is *ss* X *Ss* → 1/2 *Ss* and 1/2 *ss*. The cystic fibrosis cross is *Cc* X *Cc* → 1/4 *CC*, 1/2 *Cc*, 1/4 *cc*. They want to know about those just like the mother. Mother is a female with *Ss* and *Cc* found by multiplying 1/2 X 1/4 X 1/2 = 1/16.

3-13. This problem uses both the Multiplication Rule and the Addition Rule. If numerals occur in equal frequency, there must be 1/10 of each since there are 10 numerals. To be a 23, you have to take up a 2 and then a 3. So 1/10 for the 2 and 1/0 for the 3. 1/100 is the chance for 23. Similarly, you have to pick up 4 twice in a row to get 44. That's 1/10 X 1/10 = 1/100. So now what is the probability of picking up a 23 and a 44. It's the probability of each added together or 2/100 = 1/50.

3-14. Diagram the cross: Ii Cc BW   X   Ii Cc BW

Now consider each of the three monohybrid crosses:

Ii X *Ii*	*Cc* X *Cc*	*BW* X *BW*
1/4 *II*, 1/2 *Ii*, and 1/4 *ii*	1/4 *CC* 1/2 *Cc* 1/4 *cc*	1/4 *BB*, 1/2 *BW*, 1/4 *WW*

Those that are not ii Cc or ii CC are white. ii Cc is 1/4 X 1/2 = 1/8, and ii CC is 1/4 X 1/4 = 1/16. 3/16 will have color. 1 – 3/16 = 13/16 are white. Among the 3/16, the black color is produced 1/4 of the time. So 3/16 X 1/4 = 3/64 black. The blue color is found in 1/2 the offspring with color. So 3/16 X 1/2 = 3/32 blue.

The ratio is thus 52/64 white : 3/64 black : 6/ 64 blue : 3/64 splashed white.

3-17. For this problem we are looking at two objects in each group, and the possibilities are two heavy tubes (call it *HH*), two light tubes (*LL*) or one of each (obtained two ways as *HL* or *LH*). The binomial expansion will work on this binomial with n = 2. $(p + q)^2 = 1$, and the expanded binomial is $(p^2 + 2pq + q^2) = 1$. Let *p* be the frequency of heavy tubes, and *q* the frequency of light tubes. p = 8/20 = 0.4 and q = 12/20 = 0.6.

So the probability of two heavy tubes being picked up is $0.4^2 = 0.16$.
The probability of two light tubes is $0.6^2 = 0.36$
The probability of one heavy and one light tube being picked up is 2 (0.4 X 0.6) = 0.48.
Check to be sure 0.16 + 0.36 + 0.48 equals 1.0.

3-18. 3 p^2q is the second term: Moving 1 to the right makes the exponents pq^2. The coefficient is found in this way. 3 X 2 = 6. Then 6 /2 = 3.
28 p^2q^6 is the seventh term. Moving to the right makes the exponents change to pq^7. The coefficient is found in this way. 2 X 28 = 56. 56 / 7 = 8.
25 p^4q^3 is the fourth term. Moving to the right makes the exponents p^3q^4. The coefficient is found in this way. 25 X4 =100. Then 100/4 = 25.

$120\ p^7q^3$ is the fourth term. Moving to the right makes the exponents change to p^6q^4. The coefficient is found in this way: $120 \times 7 = 840$. Then $840/4 = 210$.

3-19. Families of six could have all six boys or five boys and a girl, or four boys and two girls or three boys and three girls or two boys and four girls or one boys and five girls or six6 girls. Set out an expansion of this binomial: $(p + q)^7 = 1$. Expanded it is this:

$$p^6 + 6p^5q + 15\,p^4q^2 + 20p^3q^3 + 15p^2q^4 + 6pq^5 + q^6$$

The sex ratio is 1:1, so $p = 0.5$ and $q = 0.5$. Thus each set of p and q are the same value at $(0.5)^6 = .015625$. The only number that varies is the coefficient. Clearly the most common six-child family is the one with three boys and three girls. Their term in the expanded binomial has a coefficient of 20, so it will have the highest numbers of families.

3-20. The probability of a woman sitting at a table is 0.40, and of a man is 0.60. Since they will be sitting in tables of four, we can do this as an expansion of the binomial $(p + q)^4 = p^4 + 4p^3q + 6p^2q^2 + 4\,pq^3 + q^4$.

There are five ways for people to sit, two of which are tables with only men or only women. The question asks the proportion of the tables with both. So omit those people (men) and q^4 (only women). What is left is the answer to the question: $p^4 = 0.6$

Probability of an all male table is $0.6^4 = 0.1296$. Probability of an all women table is $0.4^6 = 0.0256$. Added together, the probability of tables with only one gender is $.1296 + 0.0256 = 0.1552$ or approximately 16%. Then the other tables with both men and women sitting at them is $1 - .1552 = 0.8448$, or approximately 84%.

3-29 Here are tables for each of the chi-square calculations:

A

O Observed	E Expected	(O – E)	(O – E)^2	(O – E)^2/E
115	122	-7	49	0.40
129	122	7	49	0.40
			Total X^2	**0.80**
Degrees of Freedom	Probability	Accept Hypothesis?		
1	.50>p>.30	Yes		

B

O Observed	E Expected	(O – E)	(O – E)^2	(O – E)^2/E
69	58.5	10.5	110.25	1.88
115	117	-2	4	0.034
50	58.85	-8.5	72.25	1.23
Degrees of Freedom	Probability	Accept Hypothesis?	Total X^2	**3.14**
Degrees of Freedom	Probability			
2	Nearly 0.30	Yes		

C

O Observed	E Expected	(O – E)	(O – E)^2	(O – E)^2/E
68	62.25	5.75	31.625	0.508
110	130.5	-20.5	420.25	3.22
83	62.25	20.75	430.56	6.89
Degrees of Freedom	Probability	Accept Hypothesis?	Total X^2	**10.618**
2	0.01>p>0.001	No		

D

O Observed	E Expected	(O – E)	(O – E)^2	(O – E)^2/E
42	37.5	4.5	20.25	0.54
80	75	5	25	0.333
28	37.5	9.5	90.25	2.417
Degrees of Freedom	Probability	Accept Hypothesis?	Total X^2	3.290
2	Close to 0.2	Yes		

3-30 Here is a table showing all the calculations:

O	E		(O – E)	(O – E)^2	(O – E)^2/2
20	1	30	-10	100	3.333
65	2	60	5	25	0.417
40	1	30	10	100	3.333
53	2	60	-7	49	0.817
100	4	120	-20	400	3.333
71	2	60	11	121	2.017
37	1	30	7	49	1.633
60	2	60	0	0	0
34	1	30	4	16	0.533
				Total	15.416

The degrees of freedom associated with this problem is 9 – 1 = 8. The probability associated with this chi-square is between 0.1 and 0.5. So if you had set your probability before you started at 0.1, you would accept the hypothesis, but if you set it at 0.05, you would reject your hypothesis.

3-31 A. Tall X short Observe F$_2$ 787 tall : 277 short Expect 798 : 266 X^2 = 0.607. Associated Probability: 0.5>p>0.3. Accept the hypothesis.

B. Green seeds X Yellow Seeds F$_2$ 6032 yellow : 2001 green Expect 6024.75 : 2008.25. X^2 = 0.035. 70>p>50. Accept the hypothesis.

C. Yellow round seeds X Green wrinkled seeds | All F$_1$ Round and wrinkled
Cross two F$_1$ and get 315 yellow round: 101 yellow wrinkled: 108 green round : 32 green wrinkled. | The total number divided by the smallest number is 556/32 = 17.4. This is close to 16, so a good possibility is that this is a 9:3:3:1 ratio from the F$_2$ of a dihybrid cross. Calculate a chi-square. It is 0.467. Associated probability with three degrees of freedom is 0.95>p>0.90. The hypothesis is confirmed.

3-32 A. Addition of probabilities B. Chi-square test C. Expansion of a binomial
D. Multiplication of probabilities

CHAPTER 4 - MORE COMPLEX SITUATIONS

4-3. A) *Ss* is intermediate in phenotype between *SS* and *ss*, so this is a case of incomplete dominance.

B) A 3:1 ratio means the black homozygote and the heterozygote have the same phenotype. This is a case of dominance.

C) *D'D* has a phenotypic effect between *DD* and *D'D'*, so this is a case of incomplete dominance.

D) The heterozygote has both phenotypes, M and N, so this is a case of codominance.

E) If we consider lethal effect as the phenotype, the heterozygote and the wild type homozygote are the same. Thus, this is an example of dominance.

4-4.   The results will be the same. The F_1 radishes will still be heterozygous for both loci.

4-5.   We can use the Product Rule on this one. Consider the cross as two separate crosses, $C^R C^W$ X $C^R C^W$ and Hh X Hh.

The proportion of calves that will have horns and are roan is 1/8: 1/4 (those with horns) X 1/2 (those that are roan) or 1/8.

The proportion of calves that will lack horns and are white is 3/4 X 1/4, or 3/16.

*4-6.   The heterozygotes can be distinguished from both homozygotes, so neither allele is dominant. The heterozygote shows a phenotype with a less drastic effect than the p2, so this is a case of incomplete dominance.

4-10. For an ii (Type O) child to appear among the offspring, each parent must contribute an i allele. For AB to appear among the offspring, one parent must contribute an I^A and one must contribute an I^B. Therefore, one parent must be $I^A i$ and one must be $I^B i$. The cross is thus the same as in Problem 9–7, and Type O offspring will occur 1/4 of the time.

4-11. P: Cc^{ch} X $c^h c$
            Normal   Himalayan

Gametes: 1/2 C 1/2 c^{ch} 1/2 c^h 1/2 c
Punnett square:

	C	c^{ch}
c	Cc normal	$c^{ch} c$ chinchilla
c^h	$C c^h$ normal	$c^{ch} c^h$ light gray

Such a cross would result in 1/2 normal progeny, 1/4 chinchilla and 1/4 light gray.

4-12.   A) A Holstein must be ss since s is recessive to all the other alleles. So two ss cattle could only produce ss offspring. Holsteins can only produce Holsteins; they could not produce Herefords.

B) Two Herefords could each be $s^h s$. Then:

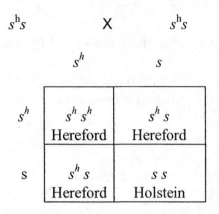

$s^h s$ X $s^h s$

	s^h	s
s^h	$s^h s^h$ Hereford	$s^h s$ Hereford
s	$s^h s$ Hereford	$s\,s$ Holstein

So Herefords could produce Holsteins.

C) Dutch belts will be $S\_$, where the blank could be any of the four alleles. Holsteins are ss, so for Dutch belts to have Holstein offspring, both parents would have to be Ss. Solid color offspring could either be $s^c s^c$ or $s^c s$. At least one parent would have to be Ss^c. So both solid color and Holstein calves could not result from mating the same parents. These are multiple alleles of *one* locus, and no parent can have more than two alleles.

*4-13. A) The three alleles can be combined in six ways: $L^N L^N$, $L^N L^H$, $L^N L^L$, $L^H L^H$, $L^H L^L$, and $L^L L^L$.

B) All genotypes with an L^H allele have the same phenotype, since that allele is dominant to the other two. L^L is dominant to L^N, so $L^L L^L$ and $L^L L^N$ have the same phenotype. Finally, $L^N L^N$ has a unique phenotype. So there will be three phenotypes.

4-16. When we cross two double heterozygotes, *Aa Bb*, for example, we get this ratio:

        9 *A_ B_*
        3 *aa B_*
        3 *A_ bb*
        1 *aa bb*

When we do a test cross of *Aa Bb* to *aa bb*, the *Aa Bb* parent produces four gamete types in equal proportion: *A B, a B, A b,* and *a b*. So four classes of offspring occur in equal proportion:

        1 *Aa Bb*
        1 *aa Bb*
        1 *Aa bb*

1 *aa bb*

The modifications come from gene interactions that result in more than one of the classes having the same phenotype.

A) A 12:3:1 ratio is obtained from the 9:3:3:1 ratio in this way:

9 A_ B_ | These two genotypes have the same phenotype.
3 A_ bb | Allele *A* is epistatic to the B–locus.
3 aa B_
1 aa bb

So the modified 1:1:1:1 ratio is

1 Aa Bb | These two genotypes would have the same phenotype,
1 Aa bb | so the ratio is 2:1:1.
1 aa Bb
1 aa bb

B) A 9:7 ratio is obtained in this way:

9 Aa Bb
3 Aa bb | These three genotypes would have the same phenotype
3 aa Bb | if a dominant allele were required at both loci to produce
1 aa bb | one of the phenotypes.

So the modified 1:1:1:1 ratio is:

1 Aa Bb
1 Aa bb | These three genotypes have the same phenotype, so the
1 aa Bb | modified ratio is 1:3.
1 aa bb

C) A 15:1 ratio is obtained in this way:

9 A_ B_ | These three classes would have the same phenotype if one
3 Aa_ bb | of the phenotypes required that both loci be homozygous
3 aa Bb | recessive.
1 aa bb

So the modified 1:1:1:1 ratio is:

1 Aa Bb | These three classes have the same phenotype, so the
1 Aa bb | modified ratio is 3:1.
1 aa Bb
1 aa bb

4-17. The problem involves a gene that is not quite completely epistatic. Allele *b* suppresses, but does not eliminate, the stripe produced by allele *S*.

P: *BB SS*   X   *bb ss*

F_1:   *All Bb Ss*   tan with a stripe

F_2:   *B_ S_*   tan with a stripe   9/16
             B_ ss   tan with no stripe or spot   3/16
             *bb S_*   brown with a spot   3/16
             bb ss   brown with no stripe or spot   1/16

Since epistasis is not complete, it's the same old double heterozygote problem with a 9:3:3:1 ratio. This emphasizes that the segregation and assortment of genes is not changed by epistasis. The only new issue is remembering how the genes interact to produce phenotypes.

4-18. Diagram what you know from how the problem is stated.

P: black   X   hairless

F_1:   All black

F_2:   Black 100   Hairless 43   Brown 30
The F_2 ratio is close to a 9:4:3 which is a modified 9:3:3:1. So the F_1 must be double heterozygotes. Let *B* be black and *b* brown. Let *H* be hairy and *h* be hairless. Then the F_1 and F_2 are:

F_1:   All black   *Bb Hh*

F_2:   Black 100   *B_ H_* 9/16
             Hairless 43   *B_ hh* 3/16 and *bb hh* 1/16 = 1/4 (4/16)
             Brown 30   *bb H_* 3/16

The parents must have been Black *BB HH* and Hairless *bb hh*.

4-19. As always, a cross between two double heterozygotes will result in a ratio of

           9 *A_ B_*
           3 *aa B_*
           3 *A_ bb*
           1 *aa bb*

The problem indicates that at least one dominant allele must be present at each locus to get black spines, so only the *A_ B_* class will be expected to have black spines. This becomes a 9 black: 7 white ratio. Of 1000 plants, 9/16 or 563 will be expected to have black spines and 7/16 or 438 will be expected to have white spines.

*4-20. One locus   Two loci

N2 N2   X   *n2 n2*   *N2 N2   N3 N3*   X   *n2 n2 n3 n3*

F_1:   *N2 n2*   *N2 n2   N3 n3*
         All with fuzz and fibers   All with fuzz and fibers

       N2 n2 X *N2 n2*   *N2 n2   N3 n3* X *N2 n2   N3 n3*

Prepare Punnett squares for each cross

	N2	*n2*
N2	*N2 N2*	*N2 n2*
n2	*N2 n2*	*n2 n2*

	N2 N3	*N2 n3*	*n2 N3*	*n2 n3*
N2 N3	*N2 N2* *N3 N3*	*N2 N2* *N3 n3*	*N2 n2* *N3 N3*	*N2 n2* *N3 n3*
N2 n3	*N2 N2* *N3 n3*	*N2 N2* *n3 n3*	*N2 n2* *N3 n3*	*N2 n2* *n3 n3*
n2 N3	*N2 n2* *N3 N3*	*N2 n2* *N3 n3*	*n2 n2* *N3 N3*	*n2 n2* *N3 n3*
n2 n3	*N2 n2* *N3 n3*	*N2 n2* *n3 n3*	*n2 n2* *N3 n3*	*n2 n2* *n3 n3*

One locus model gives three with fuzz and fiber and one fuzzless (the bottom right cell *n2 n2*).

Two locus model gives 15 with both fuzz and fiber and one fuzzless (the bottom right cell *n2 n2 n3 n3*).

Geneticists who did this cross got a ratio of 316 with fuzz and fiber to 25 fuzzless. That is consistent with the two locus model. We expect 1/16 or 22.7 fuzzless, and we got very close with 25 fuzzless.

A back cross of the F_1 to strain 143 is one of these two crosses:

         One-locus model   Two-locus model

     N2 n2 X *n2 n2*   *N2 n2 N3 n3* X *n2 n2 n3 n3*

Now do a Punnett square for each of these crosses.

| *N2* | *n2* | | *N2 N3* | *N2 n3* | *N2 N3* | *n2 n3* |

n2	*N2 n2*	*n2 n2*

	n2 n3	*N2 n2*	*N2 n2*	*n2 n2*	*n2 n2*
		N3 n3	*n3 n3*	*N3 n3*	*n3 n3*

If the one-locus model is
correct, there should be a
1:1 ratio.

If the two-locus model is
correct, there should be a
3:1 ratio.

*4-21. If a recessive allele at one locus were the basis of resistance, the cross would have been

$$RR \quad X \quad rr$$

F_1: all *Rr*

F_2: 3 *R* __ and 1 *rr* But the resistant class is nowhere near 1/4 of the progeny.

If two loci were involved, each with a dominant and recessive allele, the cross would have been

$$R\ R1\ R2\ R2 \quad X \quad r1\ r1\ r2\ r2$$

F_1: all *R1 r1 R2 r2*

F_2: 9/16 *R1* __ *R2* __, 3/16 *R1* __ *r2 r2*, 3/16 *r1 r1 R2* __ and 1/16 *r1 r1 r2 r2*

(Convince yourself of that with a Punnett square.) This fits the data fairly well. This would be a modified 9:3:3:1 ratio modified to 15:1 with resistance arising only when both loci are homozygous recessive and making up 1/16 of the total. The expected ratio would be 1068 sensitive and 71 resistant, fairly close to the 1088 sensitive and 51 actually observed.

CHAPTER 5 - SEX LINKAGE

*5-2. A) The spermatogonia are in males so X Y, and the allele is on the Y so X Y[b].
    B) Sperm has only the Y-chromosome with the *b* on it. Y[b]

5-5. If the mother is heterozygous, she will have normal color vision. Half of her offspring will receive her X-chromosome, which includes the color-blindness allele. Each son who receives that X will be color-blind because that will be their only X-chromosome. Daughters who receive the color-blindness allele from the mother will receive a normal X from the father, so they will not be color-blind.

5-6. $Vv\ X^f Y$ X $Vv\ X^F X^f$
        Male   Female

	Normal wings Forked bristles	Normal wings Normal bristles
Gametes:	1/4 $V\text{X}^f$	1/4 $V\text{X}^f$
	1/4 $v\text{X}^f$	1/4 $v\text{X}^f$
	1/4 $V\text{Y}$	1/4 $V\text{X}^F$
	1/4 $v\text{Y}$	1/4 $v\text{X}^F$

	$V\text{X}^f$	$v\text{X}^f$	$V\text{Y}$	$v\text{Y}$
$V\text{X}^f$	$VV\,\text{X}^f\text{X}^f$ Female Normal wings Forked bristles	$Vv\,\text{X}^f\text{X}^f$ Female Normal wings Forked bristles	$VV\,\text{X}^f\text{Y}$ Male Normal wings Forked bristles	$Vv\,\text{X}^f\text{Y}$ Male Normal wings Forked bristles
$v\text{X}^f$	$Vv\,\text{X}^f\text{X}^f$ Female Normal wings Forked bristles	$vv\,\text{X}^f\text{X}^f$ Female Vestigial wings Forked bristles	$Vv\,\text{X}^f\text{Y}$ Male Normal wings Forked bristles	$vv\,\text{X}^f\text{Y}$ Male Vestigial wings Forked bristles
$V\text{X}^F$	$VV\,\text{X}^F\text{X}^f$ Female Normal wings Normal bristles	$Vv\,\text{X}^F\text{X}^f$ Female Normal wings Normal bristles	$VV\,\text{X}^F\text{Y}$ Male Normal wings Normal bristles	$Vv\,\text{X}^F\text{Y}$ Male Normal wings Normal bristles
$v\text{X}^F$	$Vv\,\text{X}^F\text{X}^f$ Female Normal wings Normal bristles	$vv\,\text{X}^F\text{X}^f$ Female Vestigial wings Normal bristles	$Vv\,\text{X}^F\text{Y}$ Male Normal wings Normal bristles	$vv\,\text{X}^F\text{Y}$ Male Vestigial wings Normal bristles

Among the males, the probability is that 6/16 will have normal wings and 2/16 will have vestigial wings, for a 3:1 ratio. Also, 4/16 should have normal bristles and 4/16 should have forked bristles for a ratio of 1:1.

Among the females, there is also a 3:1 ratio of normal winged to vestigial and a 1:1 ratio of forked to normal bristles.

Of 400 flies, 200 are expected to be females, and of those 3/8 or 75 would be expected to have both normal wings and normal bristles, 3/8 or 75 will have normal wings and forked bristles, 1/8 or 25 will have vestigial wings and normal bristles, and 1/8 or 25 will have vestigial wings and forked bristles.

Similarly, the males are expected to include 75 normal, normal; 75 normal, forked; 25 vestigial, normal; and 25 vestigial, forked.

5-7. If the gene is X-linked, then Alexis must have been $\text{X}^h\,\text{Y}$, and he must have received that X^h from his mother. His mother and Alice must both have been heterozygotes since neither had hemophilia. It is most likely that Alexandra and Alice inherited the gene from a common ancestor.

*5-8. The allele is on the Y-chromosome. If a person has the allele he is XY, and his mate must be XX. So the mate could not have the allele b.

5-11. The difference in reciprocal crosses suggests X-linkage. Let B symbolize the Bar allele and b symbolize the normal. Put the alleles on the X-chromosome and the results are explained.

P: $X^b X^b$ X $X^B Y$ $X^B X^B$ X $X^b Y$
    Female   Male   Female   Male
    Normal   Bar   Bar   Normal

F_1: Males all $X^b Y$ (normal)   Males $X^B Y$ (bar)
     Females $X^B X^b$ (bar)   Females $X^B X^b$ (bar)

5-12. To decide if body color is X-linked, just look at the results for body color. The cross reduces to:

P: black-bodied male   X   yellow-bodied female

F_1:   all yellow-bodied (yellow must be dominant to black)

F_2:   males: 3 yellow : 1 black   females: 3 yellow : 1 black

Compare the results with and without X-linkage:

       If autosomal: If X-linked:

P: bb X BB $X^b Y$ X $X^B X^B$

F_1:   all Bb (yellow)   Males: $X^B Y$ (yellow)
                            Females: $X^B X^b$ (yellow)

F_2:   1 BB: (yellow)   Males: 1 $X^B Y$ (yellow) : 1 $X^b Y$ (black)
        2 Bb: (yellow)   Females: 1 $X^B X^B$: 1 $X^B X^b$ (all yellow)
        1 bb: (black)

If the body color gene were X-linked, we would see black bodies only among the F_2 males, but we see a 3:1 ratio with each sex. Therefore, the body color gene is not on the X-chromosome.

5-13. Look at the characters one at a time.

1. Long crossed with short gives a ratio of 1:1 (with a 1:1 sex ratio). We can therefore assume that this locus is not X-linked and that one parent is a heterozygote (*Ll*) and one is a homozygote recessive (*ll*).

2. Active crossed with sluggish results in all active progeny regardless of sex. It is a result of a mating (*AA* X *aa*) where active (*A*) is dominant to sluggish (*a*).

3. Purple males mated with yellow females results in yellow males and purple females. Color segregates by sex, so we can deduce that the color locus is X-linked. Purple (X^P) is X-linked dominant over yellow (X^P). The cross is $X^P X^P$ X $X^P Y$.

*5-14. An allele is said to be dominant when it has the same expression when heterozygous as when homozygous. An individual can ordinarily not have more than one Y. So the individual can never be either homozygous or heterozygous.

CHAPTER 6 - PEDIGREES

6-3. Here is the pedigree:

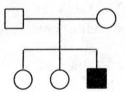

6-4. Assuming "attached earlobes" is the character under study, shade the symbols of those with attached earlobes. The pedigree:

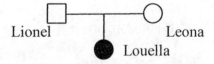

Lionel   Leona

Louella

6-5. Here is the pedigree with polydactyly shaded in:

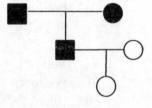

6-10. Free earlobes are dominant to attached earlobes. Both Lionel and Leona have free earlobes and yet they gave birth to a child with attached earlobes. This is a sign that attached earlobe is a recessive allele.

6-11. The man must be heterozygous. The son by the second marriage has pituitary dwarfism, a recessive character. He must have received a recessive allele from each parent. We know he got one from his homozygous mother, but he also must have a recessive allele from his father. Since the father is normal, he must also have a dominant allele, and thus is heterozygous.

6-12. Based on the data, you cannot determine whether polydactyly is dominant or recessive. The first mating in which two affected individuals had an affected child, could happen if both parents were homozygous recessives, or if they each had at least one dominant allele. If the condition were dominant, one or both parents could be heterozygous dominant, and they could have a heterozygous child. Then the child in the third generation could have received a recessive allele from each parent. Or, if the condition were recessive, the unaffected mother would have given a dominant allele (no polydactyly) to her daughter. If the couple in the second generation had a second daughter with polydactyly, that would not clear up the ambiguity: If the condition is dominant, and the affected father in the second generation were heterozygous, he could give the normal allele to the first daughter and the polydactyly allele to the second. If the condition is recessive, the mother in the second generation could be heterozygous and give the polydactyly allele to one child and the normal (dominant) allele to the other. More information is needed beyond this small pedigree.

6-13. The disability is probably due to a dominant allele. Once it appears in the family, it shows up in every generation (II-V). Affected parents have both unaffected and affected children (II-1, III-2, III-3), but two unaffected parents never have affected children (I-1 and I-2; II-6 and II-7; III-9 and III-10). (We do not know if II-3 is affected because it was stillborn.)

6-14. If the disability is due to a dominant allele, parents with the disability will have unaffected children only if the parent is heterozygous, because the parent must give a recessive allele to unaffected children. Both II-1 and III-3 have at least one unaffected child and so are heterozygous. If a person with the disability (i.e., with the dominant allele) has one normal parent, then that parent must have given a recessive allele to all offspring, and the affected children will be heterozygous. III-3, III-5, IV-9 and V-2 all have a normal parent, so all must be heterozygous.

6-15. Dominant. Parent I-1 is known to be homozygous. If the allele for PRA were recessive, I-2 would have to be homozygous to express the trait. All offspring would be unaffected. All would be heterozygous since both parents would be homozygous. But I-2 could express the trait if the PRA allele were dominant. Then the cross would be *pp* X *Pp*, and all the progeny that received a dominant allele from I-2 would be affected.

Without the beagle of known genotype, the pedigree would be ambiguous. I-1 could be homozygous or heterozygous for the recessive or homozygous for the dominant. Thus, we could not tell whether I-2 was a heterozygote being crossed to a recessive (which it is) or a homozygote recessive being crossed to a heterozygote.

*6-16. Noonan syndrome could be due to a dominant allele. I-1 would thus be homozygous recessive, and both offspring would receive a dominant allele from I-2. I-2 could be either homozygous dominant or heterozygous.

Noonan syndrome could be due to a recessive allele. Then I-1 would have to be heterozygous in order to give a recessive allele to each offspring. I-2 would have to have at least one recessive allele also but could be either homozygous recessive or heterozygous.

6-19. Heterozygous. The phenotype of interest is caused by a recessive gene. It shows up in males of the next generation, so the allele comes from the female. She has the normal phenotype, however, so she must have a dominant allele also.

6-20. The condition is recessive and X-linked. II-5 and IV-2 arise from two unaffected parents. This is a good indication that the gene is recessive. All the affected individuals are males. The affected individuals in Generation IV all have mothers with an affected father. These are good signs of X-linkage.

6-21. Since the allele is recessive and X-linked, a male cannot receive the allele at issue from his father, since he gets a Y-chromosome from the father. II-5 had to receive his allele from his mother; therefore, IV-6 is wrong about the great grandfather's role in introducing the condition into the family.

*6-22. A) Bilateral Perisylvian Polymicrogyria (BPP) is probably due to an X-linked recessive. Evidence for the recessive condition comes from the appearance of BPP from two unaffected parents. It would help to have a few females in Generation III, but the results are all consistent with X-linkage. Two unaffected males have affected male offspring with females from the same family. The condition is rare enough that it is unlikely either II-1 or II-4 carries the BPP allele. If I-1 has one copy of the normal allele, and I-2 were heterozygous, then two heterozygous (unaffected) females are possible in Generation II. If the allele were not sex-linked, all four of Generation II would have to be heterozygous, an unlikely situation given one BPP per 2500 live births.

B) Bilateral Frontoparietal Polymicrogyria (BFP) is probably due to a recessive autosomal allele. It only appears in a mating of two unaffected individuals, thus establishing recessiveness. Females with BFP come from two unaffected parents. This could not happen if the allele were X-linked. Then the male parent would have to have the BFP allele in order for daughters to have two of them. III-2 does not.

C) Susceptibility to pancreatic cancer is probably an autosomal dominant. As with the preceding two cases, this susceptibility allele is relatively uncommon,` so the people marrying into the main family probably do not have the condition. Thus, if the allele were X-linked, affected males could not give rise to both affected males and affected females. This happens frequently in the pedigree, however. If the allele were a recessive, I-1 would have to be a homozygote and I-2 would most probably be a homozygous dominant. We would not expect all but one of their offspring to express the susceptibility allele. Likewise, so many affected offspring of II-1 and II-1 would not be expected. But if the susceptibility allele were dominant, that is exactly what we would expect to see.

CHAPTER 7 - LINKAGE

7-3. The cat must have received an *nH* from one parent and an *Nh* from the other parent. Since the parents are both homozygotes, they must be *nH/nH* and *Nh/Nh*.

7-4. The offspring will receive one chromosome from each parent, so the genotype is *ABc/abC*.

7-5. One mouse is homozygous for *f* (*fT/fT*), and the other mouse is homozygous for *t* (*Ft/Ft*).

***7-6.** A) Those loci that are on the same chromosome should show linkage. Those include glycogen phosphorylase and lactate dehydrogenase on Chromosome 11, enolase and very-long-chain acyl-CoA dehydrogenase on Chromosome 17, and phosphorylase kinase and phosphoglycerate kinase on Chromosome X.

B) Let the glycogen phosphorylase alleles be *G* and *g*, and the LDH alleles be *L* and *l*. The designated individual has one parent *g L/g L* and the other parent *G l/G l*. That person is thus *g L/G l*.

Let the enolase alleles be *E* and *e* and the very-long-chain acyl-CoA dehydrogenase be *V* and *v*. One parent is *E v/E v*, and the other is *e V/e V*, and the designated individual is *E v/e V*.

Let the phosphorylase kinase alleles be *P* and *p* and the phosphoglycerate kinase be *K* and *k*. The male parent may be *P k*, in which case the female parent is *p K/p K*, or the male parent may be *p K*, in which case the female parent is *P k/P k*. In either case, the female from such parents is *P k/p K*. The male in the first case is *p K*, and in the second case is *P k*.

7-9. The *AA BB CC* parent produces one kind of gamete – *ABC*. The *aa bb cc* parent also produces one kind of gamete, *abc*. The progeny would all be *AaBbCc*. This will be true whether the genes are linked or not. If the genes are linked, we'd say *ABC/abc*.

7-10. The gametes would be of two kinds if the genes are linked, just like the gametes of the two original parents: *ABC* and *abc*. So if crossed to a parent that is *aa bb cc*, the offspring will be *Aa Bb Cc* and *aa bb cc*. But since the problem specifies that all three loci are linked, the offspring should be symbolized *ABC/ABC* and *abc/abc*.

7-11. Genes are linked when they are on the same chromosome. In this case, both vermilion and miniature are on the X-chromosome. So not only are they X-linked, they're linked.

*7-12. *G l/g L*   *G l* and *g L*

E v/e V   *E v* and *e V*

Female *P k/p K*   p K *and* P k   Male *p K* just *p K*   Male *P k just P k*

7-14. In each case, there will be one parent *cw/cw*. This will only form one kind of gamete: *cw*. The types and proportions of gametes from the heterozygous parent will vary in each instance.

CW/cw   X   *cw/cw*

Genotypes of Test Cross Offspring	Male Heterozygote	Female Heterozygote	If Independent Assortment Happened
CcWw	0.5	0.4	0.25
Cc ww		0.1	0.25
ccWw		0.1	0.25
cc wv	0.5	0.4	0.25

An excess of parentals is seen in both cases of linkage, even though there is some recombination in the female.

*7-15. Parental types   S5 S13   175
                                     s 5 s13   175
  Recombinant types   S5 s 13   75
                                     s5 S13   75

*7-16.

Parental types	S5 s13	175
	s 5 S13	175
Recombinant types	s5 s 13	75
	S5 S13	75

7-19. The classes that are parental and recombinant would be switched around. The parental gametes would be *Vm* and *vM*, and the crossover gametes would be *VM* and *vm*. The proportions of parentals would still be 97%, and the proportion of crossover gametes would still be 3%.

7-20. Consider the two crosses. Remember that these are X-linked genes.

Female	*VM/VM*	X	Male	*vm/Y*		Female *vm/vm*	X	Male *VM/Y*
Gametes:	all *VM*			*vm*		all *vm*		*VM*
				Y				Y
Offspring:	Females	*VM/vm*				Females	*VM/vm*	
	Males	*VM/Y*				Males	*vm/Y*	

Because of the way X-linked inheritance works, only the cross on the right will yield the desired genotypes.

*7-21. All the loci on one chromosome make a linkage group. Genes on different chromosomes show independent assortment. Since *Lotus japonicus* has six linkage groups, its haploid number is six and the diploid number is 12. Likewise, *Antirrhinum majus* with eight linkage groups probably has a diploid number of 16.

CHAPTER 8 - LINKAGE MAPS

8-3. The 447 recombinants (i.e., the ones in which the alleles are not in the same combinations as they are in the parents) are 34.7% of the progeny, so the map distance is 34.7.

8-4. In this case, the wild type and white miniatures would be the recombinant types. There are 34.7 map units between white and miniature, so 34.7% of the progeny would be these recombinant types.

*8-5. A) If the distance from N1 to the end of the chromosome is 32 map units, and the distance of C2 from the end of the chromosome is 23 map units, then the distance between N1 and C2 should be 9 map units. Since map units are the % recombination, the recombination between N1 and C2 is 9% or 0.09.

B) 15 – 7.5 = 7.5 map units. The recombination between L and P is 7.5% or 0.075.

8-8.

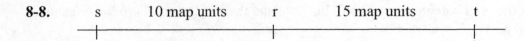

s   10 map units   r   15 map units   t

8-9. Distance between the antenna and eye genes

Long gold and short silver are the parental types, so long silver and short gold are the recombinants. 198 + 202/1000 is 400/1000 or 0.40. That is 40 map units.

Distance between the antenna and bristle genes

Long close and short fair are the recombinants in this case. There are 50 + 50 of the recombinants, and 50 + 50/1000 is 0.10 or 10 map units.

Distance between eye and bristle genes

Gold close and silver far are the recombinants. Therefore 228 + 224/ 1000 is 0.452, and the map distance is 45.2 map units.

eye   40 map units   antenna   10 map units   bristles

*8-10.

Parentals	D2	*Pax6*	Ag
	d2	*Pax6*	ag
Recombinants			
Between D2 and *Pax6*	D2	*Pax6*	ag
1.41%	d2	*Pax6*	Ag
Between *Pax6* and Ag	D2	*Pax6*	ax
25.3	d2	*Pax6*	Ax

8-13. In the data for problem 8-11, the most frequent classes are white, full, waxy and colored, shrunken, starchy. These are the parental classes that have not recombined. The least frequent classes are white, shrunken, waxy and colored, full, starchy. These are the double crossover classes. The alleles of the middle locus will change places relative to the other two. In this case white and waxy remain together and so do colored and starchy. The shrunken/full locus trades places between the parental and double crossover classes, so that is the middle gene, and the order is

white/colored   shrunken/full   waxy/starchy

8-14. First arrange the classes of offspring form largest number in the class to smallest.

RsT   8820 + 9215 = 18035
rSt   9111 + 8901 = 18012

rst   1234 + 1306 = 2540
RST   1243 + 1187 = 2420

RSt   926 + 889 = 1815
rsT   942 + 907 = 1849

Rst   8 + 7 = 15
rST   5 + 8 = 13

The most frequent classes are parental classes, and the least frequent are due to double crossovers. In the double crossover class the alleles of the muffle locus will have changed places relative to the outside loci.

In this case the two parental genotypes are *RsT* and *rSt*. The Doubles are *Rst* and *rST*. Note that the combinations *Rs* and *rS* do not change. The *T/t* locus changes places. So the order of these loci is   R/r T/t S/s.

8-17. A) The most common types are *MNP* and *mnp,* so those are probably the parental (non-crossover) types. So the heterozygous parent would be *MNP/mnp*.

      B) The heterozygous parent's two homozygous parents are *MNP/MNP* and *mnp/mnp*.

      C) Divide the problem into three pair-wise problems

MN	409		*MP*	422		*NP*	352
Mn	90		*Mp*	77		*Np*	127
mN	70		*mP*	83		*nP*	153
mn	431		*mp*	418		*np*	368

Crossovers	Crossovers	Crossovers
(90+70/1000) X 100%	(27+83/1000) X 100%	(127+153/1000) = 100%
16% or 16 map units	16% or 16 map units	28% or 28 map units

The outside genes are N and P, so the map is:

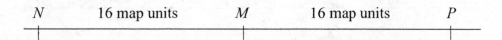

   N   16 map units   *M*   16 map units   *P*

8-18. Arrange the data into three problems of two genes each.

DP	2130	*PE*	2452	*DE*	1991
dp	2090	*pe*	2380	*de*	1959
dP	550	*pE*	220	*dE*	681
Dp	510	*Pe*	228	*De*	649

Crossovers

(550+510/5280) X 100%

20.1 % or 20 map units

Crossovers

(220+228/5280) X 100%

8.5% or 8.5 map units

Crossovers

(681+619/5280) X 100%

25% or 25 map units

The outside genes are *D* and *E*, so the map is

| *D* | 20 map units | *P* | 8.5 map units | *E* |

*8-19. A) If there were no linkage, the result of a three-point test cross should be a 1:1:1:1:1:1:1:1 ratio. There is an excess of some classes relative to others, so there does appear to be linkage. To see which ones are linked, group the data as three crosses of two loci each.

H	H	H	h
C	c	C	c
633	13	18	660

H	H	h	h
D^R	D^s	D^R	D^s
368	278	298	379

C	C	c	c
D^R	D^s	D^R	D^s
365	290	301	371

There is a clear excess of parental types when H and C are observed. There may be slight excesses in the other two cases as well.

B) Making the map.

   1. Arrange the data into reciprocal classes

Parentals	Parentals	CO I	CO I	CO II	CO II	Doubles	Doubles
H D^R C	h D^s c	H D^s C	h D^R C	H D^R C	h D^s C	H D^s C	h D^R C
359	367	274	292	9	12	4	6

Compare the parentals and the doubles: The H alleles have shifted position relative to D and C. So the order is C H D. Notice in the pair-wise groupings that the smallest number of recombinants was between h and c, so the map is like this.

C	II	H		I		D

Frequency of recombinants between C and H are (C.O.II + Doubles)/total

$(9 + 12 + 4 + 6)/1323 = 31/1323 = 0.024$ or 2.4 %

Frequency of recombinants between H and D are (C.O. I + Doubles)/total

$(274 + 292 + 4 + 6)/1323 = 576/1323 = 0.434$ or 43.4 %

C	2.4	H	43.5		D

CHAPTER 9 - CHROMOSOMES

*9-3. Here are rough diagrams based on the descriptions.

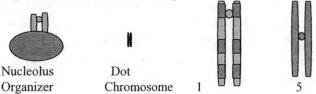

Nucleolus Organizer   Dot Chromosome   1   5

9-5. Chromosomes have two chromatids when they condense in prophase. Then when they split apart at anaphase, each chromosome has one chromatid. So answer this question simply by counting the number of chromatids per chromosome.

A) Anaphase   B) Prophase   C) Anaphase

 9-6. The function of mitosis is to keep the chromosomes constant through many cell divisions. The genes are on the chromosomes. So if all of our cells descended from a zygote by repeated mitosis, they should all have the same genes.

*9-7. The frog with a diploid number of 22 has 11 pairs of chromosomes. Subtract the six short and one long chromosome form the haploid number and you have 4 left. These are the intermediate ones. At mitotic anaphase, one of each chromosome will get to each pole. That number is 4.

The other frog species has 24 chromosomes in their diploid set. That is the number that will be in a prophase cell.

*9-10 Good condensed chromosomes that are easy to see make the best karyotypes. Those come from chromosomes during mitosis, often metaphase. If a karyotype is made from chromosomes before anaphase, each chromosome will have one centromere. The house mouse will have 40 chromosomes, thus 40 centromeres; and the spiny mouse will have 26 chromosomes, thus 26 centromeres in its karyotype.

9-12.   A) At Anaphase II of meiosis, one chromatid per chromosome and a haploid number of chromosomes would go to each pole in a diploid. Since a tetraploid has four sets, two chromatids will go to each pole, but still one chromatid per chromosome. See two of the particular chromatid.

B) In Anaphase I in a diploid, two chromatids per chromosome, but a haploid set of chromosomes goes to each pole. So the number of this chromatid per pole is two.

C) In anaphase of mitosis, one of every complete set of chromosome goes to each pole with one chromatid. A hexaploid has six complete sets of chromosomes, so we would see six chromatids of this type going to each pole.

9-13.   The key to meiosis is synapsis. Each chromosome pairs with its homologue, but in pentaploids or triploids, such pairing cannot occur with all chromosomes since there is an odd number of chromosome sets. A triploid might get two sets synapsed, but not the third, and a pentaploid might get four sets synapsed, but not the fifth. Therefore, the extra chromosomes will not be assorted systematically to the poles at Anaphase I.   At mitosis each individual chromosome splits, and one chromatid of every chromosome travels to each pole. This will happen just as easily in a triploid as a diploid since chromosome pairing is not part of mitosis.

*9-14.   The pairs of spots probably reveal centromeres on synapsed homologous chromosomes. There is a diploid number of centromeres for each contributing species, and the total of both is the diploid number of A. suecica. Arabidopsis arenosa has a diploid number of 2 X 8 = 16, A. thaliana has 2 X 5 = 10, and the allopolyploid A. suecica has 2 X 13 = 26 chromosomes in a diploid set.

9-16. A) One purple and one of every other color too.
B) Two red chromosomes, each with two chromatids, is four red chromatids.
C) The diploid stage undergoes meiosis.
D) Meiosis requires synapsis. A haploid set of chromosomes consists of unpaired chromosomes. Since there are no homologues for them to pair with, chromosomes of a haploid set cannot undergo meiosis.
E) At the end of meiosis, a complete haploid set will be in each nucleus. So five colors will be seen.
F) If there are five pairs of chromosomes, and each pair differs in the lightness of each stain, there will be 2^5 or 32 different possible combinations. Let R, G, Y, B and P stand

for the five colors of chromosomes that stain darkly, and r, g, y, b and p for the chromosomes that stain darkly. In a diploid set there are these chromosomes:

Rr Gg Yy Bb and Pp

Use the branch or list method to sort out all possible combinations:

R G Y B P	r G Y B P
R G Y B p	r G Y B p
R G Y b P	r G Y b P
R G Y b p	r G Y b p
R G y B P	r G y B P
R G y B p	r G y B p
R G y b P	r G y b P
R G y b p	r G y b p
R g Y B P	r g Y B P
R g Y B p	r g Y B p
R g Y b P	r g Y b P
R g Y b p	r g Y b p
R g y B P	r g y B P
R g y B p	r g y B p
R g y b P	r g y b P
R g y b p	r g y b p

9-17. Chromosome 11 at synapsis

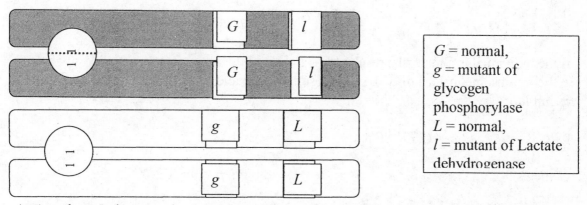

G = normal,
g = mutant of glycogen phosphorylase
L = normal,
l = mutant of Lactate dehydrogenase

At Anaphase I, the grey chromosome will go to one pole, the white chromosome to the other.

Likewise, the X chromosome will have two loci on it, and will look like this in females if the loci are heterozygous.

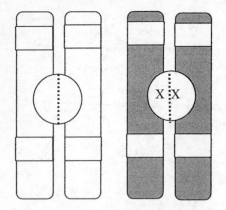

P = normal phosphoglycerate kinase
p = mutant form
K = normal phosphorylase knase
k = mutant from

They will line up in metaphase I like this, too, and thus at anaphase I the gray chromosome will either go with the gray or white chromosome 11.

If grey goes with gray , the gametes will be $G\ l\ P\ K$ and $g\ L\ p\ k$. If the gray 11 goes with the white , the gametes will be Glpk and gLPK.

Now think of this as a genetics problem without considering chromosomes.

Parents: GG ll PP KK   X   gg LL pp kk

Gametes: $G\ l\ P\ K$ $g\ L\ p\ k$

F_1: $G\ l\ /\ g\ L$ $P\ K\ /\ p\ k$

Gamete (Compare to chromosome pictures)

     $G\ l\ P\ K$ $G\ l\ p\ k$ $g\ L\ P\ K$ $g\ L\ p\ k$

If the individual is a male, the results for g and l will the same. But Males will only have 1 X chromosome instead of two, so the results with P and K will vary. Here is one example:

Parents :   GG ll PP KK   X   gg LL p k Y

Gametes : $G\ l\ P\ K$ $g\ L\ p\ k\ \ or\ \ g\ L\ Y$

F1 :   Gg Ll Pp Kk or Gg Ll P K Y

Chapter 10 - Population Genetics

10-7.   500 − 481 = 19 albinos.   19/500 = 0.038 albinos.

10-8.   Since a is recessive, the proportion of a in the Aa class cannot be known because the AA cannot be distinguished from Aa.   Neither $f_{(AA)}$, $f_{(Aa)}$ nor $f_{(a)}$ can be determined.

10-9. From Problem 10-7, we know that $aa = 0.038$ and this problem tells us that the frequency of heterozygotes is 0.396. So

	AA	Aa	aa
	0.566	0.396	0.038

Frequency of AA was determined by subtracting the frequencies of Aa and aa. All the AA and half the Aa give us the frequency of A of 0.764. For the frequency of a use all the aa and half the frequency of Aa to get 0.236. Sure enough, $0.764 + 0.236 = 1$.

10-10.

Population 1

Initial Zygotes	AA	Aa	aa
Zygote Frequencies	0.5	0.2	0.3
Resulting Gametes	A	a	
Gamete Frequency	$0.5 + 1/2 (0.2) = 0.6$	$0.3 + 1/2a(0.2) = 0.4$	
1st Generation Zygotes	AA	Aa	aa
Zygote Frequencies	0.36	0.48	0.16
Resulting Gametes	A	a	
Gamete Frequency	$0.36 + 1/2 (0.0.48) = 0.6$	$0.16 + 1/2 (0.48) = 0.4$	
2nd Generation Zygotes	AA	Aa	aa
Zygote Frequencies	$0.6^2 = 0.36$	$2 (0.6)(0.4) = 0.48$	$0.4^2 = 0.16$

Population 2

Initial Zygotes	AA	Aa	Aa
Zygote Frequencies	0.25	0.5	0.25
Resulting Gametes	A	a	
Gamete Frequency	$0.25 + 1/2 (.5) = 0.5$	$0.25 + 1/2 (0.5) = 0.5$	
1st Generation Zygotes	AA	Aa	aa
Zygote Frequencies	$0.5^2 = 0.25$	$2 (0.5)(0.5) = 0.5$	$0.5^2 = 0.25$
Resulting Gametes	A	a	
Gamete Frequency	$0.25 + 1/2 (.5) = 0.5$	$0.25 = 0.5$	
2nd Generation Zygotes	AA	Aa	aa
Zygote Frequencies	$0.5^2 = 0.25$	$2(0.25)(0.25) = 0.5$	$0.5^2 = 0.25$

Population 3

Initial Zygotes	AA	Aa	Aa
Zygote Frequencies	0.04	0.8	0.16
Resulting Gametes	A	a	
Frequency	$0.04 + 1/2 (0.8) = 0.44$	$0.16 + 1//2 (0.8) = 0.56$	
1st Generation Zygotes	AA	Aa	aa
Zygote Frequencies	$0.44^2 = 0.1936$	$2 (0.44)(0.56) = 0.4928$	$0 56^2 = 0.3136$
Resulting Gametes	A	a	
Gamete Frequencies	$0.1936 + 1/2 (.4928) = 0.44$	$0.3136 + 1/2 (4948) = 056$	
2nd Generation Zygotes	AA	Aa	aa
Zygote Frequencies	$0.44^2 = 0.1936$	$2 (0.44)(0.56) = 0.4928$	$0 56^2 = 0.3136$

10-12.

Gray GG	Gray Gg	Black gg
$25/50 = 0.5$	$20/50 = 0.4$	$5/50 = 0.1$

Gametes $\qquad G = (25 + 10)/50 = 0.7 \qquad g = (5 + 10)/50 = 0.30$

Next generation $\qquad GG \qquad\qquad\qquad Gg \qquad\qquad\qquad gg$

$\qquad\qquad\qquad\qquad 0.7^2 = 0.49 \qquad 2 \text{X} 0.7 \text{X} 0.3 = 0.42 \qquad 0.3^2 = 0.09$

Very close to equilibrium. Perhaps there has been a small sampling error. Only 50 individuals is not a very large population.

10-13. Question $\qquad\qquad\qquad\qquad\qquad$ Equilibrium?

$\qquad$ 10-4 ………………………… …………………yes

$\qquad$ 10-5 (Kidd)…………………………….. no

$\qquad$ 10-10 (1) ………………………………….no

$\qquad$ 10-10 (2) ………………………………….yes

$\qquad$ 10-10 (3) …………………………………..yes

10-14.

A) $p^2 + 2pq + q^2$

B) $p^2 + q2 + r^2 + 2\,pq + 2\,pr + 2qr$

C) $p^2 + q^2 + r^2 + s^2 + 2pq + 2pr + 2ps + 2qr + 2qs + 2rs$

10-15. Use a Punnett square to find all the possible zygote frequencies.

	p 0.1	q 0.4	r 0.2	s 0.3
p 0.1	p2 0.01	pq 0.04	pr 0.02	ps 0.03
q 0.4	pq 0.04	q^2 0.16	qr 0.08	qs 0.12
r 0.2	pr 0.02	qr 0.08	r^2 0.0.04	rs 0.06
S 0.3	ps 0.03	qs 0.12	rs 0.06	s^2 0.09

The zygote frequencies are found on the Punnett square. Those are symbols of frequencies. Now we translate those into genotypes of the zygotes:

vv (p^2)	ww (q^2)	yy (r^2)	zz (s^2)	vw $(2pq)$	vy $(2pr)$	vz $(2ps)$	wy $(2qr)$	wz $(2qs)$	yz $(2rs)$
0.01	0.16	0.04	0.09	0.08	0.04	0.06	0.16	0.24	0.12

10-16

Gamete Type	Gamete Frequency	
V	All the VV and half each of vw, vy and vz .	0.01 + 0.04 + 0.02 + 0.03
	0.1	
W	All the WW and half each of vw, yw and wz.	0.16 + 0.08 + 0.04 + 0.12
	0.4	
Y	All the yy and half each of vy, wy and zy	0.04 + 0.02 + 0.16 + 0.06
	0.28	
Z	All the zz and half each of vz, wz and yz.	0.045 + 0.03 + 0.12 + 0.06
	0.20	

Previous Gamete Frequency		Next generation of gametes
v	0.1	0.10
w	0.4	0.4
y	0.2	0.28
z	0.3	0.20

The two sets of gametes may or may not show the same frequencies. It is close enough to suggest that the two sets of gametes are from a gene pool in Hardy-Weinberg Equilibrium.

CHAPTER 11 – QUANTITATIVE GENETICS

11- 4 There is little need to construct a data table for 10 worms, none of which are the same lengths. We can, however, arrange the data in numerical order.
9, 12, 8, 6, 10, 7, 13, 4 ,2, 11 ⟶ 2, 4, 6, 7, 8, 9, 10, 11, 12 13,

Median: The sample size is 10. That means there will be five numbers on each side of the middle that puts the median in the space between 8 and 9.
Then the median is $(9 + 8)/2$ or 8.5.

11-5. Add the segments to the ones in the last problem. Three lengths, one each of 9, 10, and 11 segments turn 11-4 into a problem with a sample size three lengths greater (13 instead of 10), and with $10 + 11 + 12 = 33$. So now the sum is $(82 + 33) = 115$ and the mean $115/13 = 8.8$. The mean has increased because we added lengths all of which were greater than the previous mean.

The median is found using those data that are arranged in numerical order: 2, 4, 6, 7, 8, 9, 10, 11, 12 13. Add the three new lengths to get 2, 4, 6, 7, 8, 9, 10, 10, 11, 11, 12, 12, 13. This sequence is 13 individuals, an odd number, telling us that one of these numbers is the at the median. Thirteen divided in half is two lists of 6 each plus one in the middle (a 10).

11-6. Order the numbers of chips from small to large and then multiply the number n the first row by the corresponding number in the second row.

Number of chips	0	1	2	3	4	5	6	7
Cookies	3	3	6	6	8	6	4	2
NXC	0	3	12	18	32	30	24	14

The final total of the second row is the sample size. The sum of the values in the third row is the total observations of chocolate chips. So mean = sum of row three/sum of second row. Mean = 133/38 = 3.5 chips per cookie

11-7

Student	Test 1	Test 2	Test 3	Test 4	Mean	Median
Irwin	79	80	85	84	82	82
Lola	65	84	70	78	74.25	74
Merry	90	95	89	91	91.25	92
Wilson	95	84	89	64	80.5	86.5
Meng	83	70	70	84	77.25	83.5

11-8.   A. 1, 2, 1, 3, 1   B. 5, 6, 7, 8, 9   C. 60, 30, 25, 55, 45

In numerical order   1, 1, 1, 2, 3   5, 6, 7, 8, 9   25. 30. 45, 55, 60

A. Mean = 1.6	B. Mean = 7	C. Mean = 43
Median = 1	Median = 7	Median = 45
Variance = 3	Variance = 2.5	Variance = 232.5
Stan Dev = 1.73	St D = 1.54	St. Dev. = 15.25
Coefficient of Variation = 1.73/	C.V = 1.58	C.V.= 15.25/43 = 0.51

11-9. First make a data table showing the number of captures and the size of the captures. Then fill in the rest of the tables for calculating descriptive statistics. The order of events is important. Subtract the mean before you multiply by the number captured.

1	2	3	4	5	6	7	8	9	10	Prey size (mm)	
34	24	18	12	6	4	3	2	1	0	Number caught	Sum = Sample size = (n) = 104
34	48	54	48	30	24	21	16	9	0	Size X number	Sum = Total measures = 284
											Mean = 284/104 = 2.730

Then add more to the table to help calculate the variance and standard deviation.

-1.73	-0.73	0.27	1.27	2.27	3.27	4.27	5.3	6.77	7.73	Subtract the mean from each value
2.99	0.53	0.07	1.61	5.15	10.6	18.2	27.7	38.4	7.45	Square the difference obtained
34	24	18	12	6	4	3	2	1	0	Obtain The number at each size
101.7	12.7	1.26	19.3	30.9	42.4	55.4	54.8	38.4	0	Multiply that by the square of the difference

Sum /n-1 = 353.9 /103 = 3.44 = variance	Square root = 1.85 = standard deviation

11-10 The denominator of the standard deviation equation is n – 1. If you had a sample size of 1, the denominator would be 0. We cannot divide by 0.

11-11 Draw two graphs of the data from 11-6 one as a histogram and one as a line graph connecting the points. 2, 4, 6, 7, 8, 9, 10, 11, 1, 13

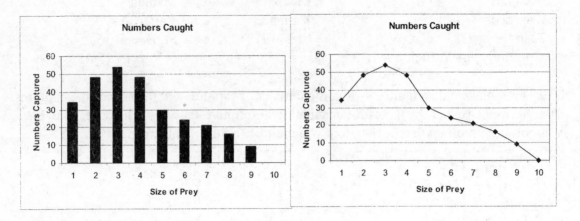

1-12 The first graph shows the raptor data with 1 cm intervals

The second graph shows the same data with 3 cm intervals.

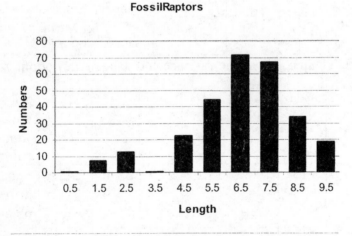

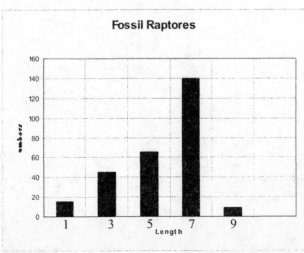

11-14. Here is the Punnet Square. Top line in each cell is the genotype, the second line is the phenotype with regard to the lethal. The third line is the size of that particular organism given the lethal gene, and the third line is what we would see if the b allele were not a recessive lethal as it is in this problem.

	AB	A b	a B	a b
AB	AA BB Live (4 + 16) =20	AA Bb Live (3 + 16) - 19	Aa BB Live (3 + 16) = 19	AaBb Live (2 + 16)= 18
A b	Aa B b Live (2 + 16) = 18	AA bb Dead	Aa Bb Live (2 + 16) 18	Aa bb Dead
a B	Aa BB Live (3 + 16) = 19	Aa Bb Live (2 + 18) = 18	Aa BB Live (3 + 16) = 19	aa Bb Live (1 + 16) = 17
a b	Aa Bb Live (2 + 16) = 18	Aa bb Dead	a a Bb Live (1 + 6)= 17	aa bb Dead

Top Row = Intervals; Second Row is number per interval

16	17	18	19	20	No lethal
1	3	7	4	1	
16	51	126	76	20	

Top Row = Interval; Second Row is number per interval

16	17	18	19	20	Lethal
1	2	5	4	1	
0	34	90	76	20	

CHAPTER 11-A COLLECTION OF REVIEW PROBLEMS

R-1.   The difference between male and female offspring ratios suggests sex linkage. The 1:1 ratio in the males suggests one locus with two alleles. The appearance of raspberry progeny from a cross of two wild-type flies suggests that raspberry is recessive to wild type.

Number of loci?   1
Number of alleles?   2
Dominant and recessive?   Wild type is dominant to raspberry.
Autosomal or X-linked?   X-linked
Independent assortment or linkage?   Not applicable
If linkage, what are the map order and map distances?   Not applicable

R-2.   The ratio in the F_2 is close to a 9:3:3:1, which suggests two independently assorting loci, each with a dominant allele for wild-type eye color. The 1/16 category should be the double homozygous recessive. So this is the cross:

P:	Scarlet	X	Brown
	ss BB		*SS bb*

F_1:	all wild type
	Ss Bb

F_2:	440 : 137 : 150 : 48	9	*S__ B__*	wild type
	Divide all by 48 to get ratio	3	*ss B__*	scarlet
	9.17 : 2.85 : 3.13 : 1 or nearly 9:3:3:13	*S__ bb*	brown
		1	*ss bb*	white

Number of loci?   2
Number of alleles?   2 at each locus
Dominant and recessive?   Wild type is dominant; scarlet and brown are recessive.
Autosomal or X-linked?   Autosomal
Independent assortment or linkage?   Independent assortment
If linkage, what are the map order and map distances? Not applicable

R-3.   The excess of two types when the categories are supposed to be equal in number suggests linkage. The excess categories should have the genotype of the original parents.

	Female	*B c/b C*	X	Male	*b c/Y*

Gametes:		Four kinds		Two kinds
	Parental	Crossovers		
	B c	*B C*		*b c*
	b C	*b c*		Y

Punnett square:

	b c	Y	Numbers	%
B c	*B c* *b c* Bar Carnation Female	*B c* Y Bar Carnation Male	470 males 500 females 970 total	48.5%
b C	*b C* *b c* Normal Normal Female	*b C* Y Normal Normal Male	465 males 475 females 940 total	47.0% 95.5% parental
BC	*B C* *b c* Bar Normal Female	*B C* Y Bar Normal Male	20 males 18 females 38 total	1.9%
b c	*B c* *b c* Normal Carnation Female	*b c* Y Normal Carnation Male	28 males 25 females 53 total	2.6% 4.5% Cross- overs

The distance between the genes in map units is equal to the % of crossover or recombinant types. In this case that is $1.9 + 2.6 = 4.5$ map units.

Number of loci? 2

Number of alleles? 2 at each locus

Dominant and recessive? Bar is dominant to normal eyes; normal eye color is dominant to carnation.

Autosomal or X-linked? The problem specifies sex-linkage.

Independent assortment or linkage? Linkage

If linkage, what are the map order and map distances? The loci are 4.5 map units apart.

R-4. The fact that wild-type parents give eyeless and purple offspring tells us that the eyeless and purple alleles are recessive to wild-type alleles. Let *e* be the eyeless allele and *E* its alternative wild type, and *p* be the purple allele and P its wild-type alternative. The test cross of each offspring is to individuals that are *ee pp*.

A 1:1:1:1 ratio comes from a test cross of a double heterozygote, and a 1:1 test cross comes from a test cross of a single heterozygote. All wild-type progeny from a test cross means that the fly being tested has no recessive alleles. So we have these four kinds of offspring being test crossed:

$$Ee \quad Pp \quad X \quad ee\,pp \quad (1:1:1:1)$$
$$Ee \quad PP \quad X \quad ee\,pp \quad (1:1)$$
$$EE \quad Pp \quad X \quad ee\,pp \quad (1:1)$$
$$EE \quad PP \quad X \quad ee\,pp \quad \text{(all wild type)}$$

How could two wild-type flies have produced these four kinds of offspring?
Since both are wild type, they are *E__ P__*. They can't both have an *e* allele or eyeless progeny would have been produced by the original mating. Likewise, no purple progeny from the original mating probably means they can't both have a *p* allele.

Either of these genotypes could be those of the original parents

Either: *Ee Pp*   X   *EE PP*

Gametes: *E P*   *e P*   *E P*
             E p   *e p*

A Punnett square shows all are wild type:

	E P	*E p*	*e P*	*e p*
E P	*EE PP*	*EE Pp*	*Ee PP*	*Ee Pp*

Or   *Ee PP*   X   *EE Pp*

Gametes: *E P*   *E P*
           e P   *E p*

A Punnett square shows all are wild type:

	E P	*E p*
E P	*EE PP*	*EE Pp*
e P	*Ee PP*	*Ee Pp*

Number of loci? 2
Number of alleles? 2 at each locus
Dominant and recessive? Eyeless and purple are recessive.
Autosomal or X-linked? Autosomal
Independent assortment or linkage? Independent assortment
If linkage, what are the map order and map distances? Not applicable

R-5. Here is the cross. Because the male and female progeny are different, we will assume sex linkage and see if it works.

P: $X^w \, X^{w'}$ X $X^W \, Y$

          heterozygous female   wild-type male

Gametes: X^w or $X^{w'}$ X^W or Y

Punnett square:

	X^W	Y
X^w	$X^W \, X^w$ Wild-type female	$X^w \, Y$ White-eyed male
$X^{w'}$	$X^W \, X^{w'}$ Wild-type female	$X^{w'} \, Y$ Apricot-eyed male

Number of loci? 1
Number of alleles? 3 (multiple alleles)
Dominant and recessive? Wild type is dominant to w and w'; w and *w'* show incomplete dominance to each other.
Autosomal or X-linked? X-linked
Independent assortment or linkage? Not applicable
If linkage, what are the map order and map distances? Not applicable

R-6. A) The woman is *Aa Rr*. With the *R* allele, she has retinitis pigmentosa.

B) Her husband has the disease but neither parent did, so he must have the recessive form due to *aa*. Parents each contributed an *a* allele. The man is *aa rr*.

Man   *aa rr*   Woman   *Aa Rr*

Gametes: all *a r*   *AR, Ar, a R, ar*

Offspring: *Aa Rr*   *Aa rr*   *aa Rr*   *aa rr*

Only *Aa rr*, or ¼ of their children will be free of retinitis pigmentosa.

R-7. A) This child is *A*, so his genotype must be I^A___.   The mother is *O*, so her genotype must be $i^o \, i^o$.

    197

Therefore, the child is $I^A\,i^O$, and the father must have supplied the I^A. Thus, the father could be Group A ($I^A\,I^A$ or $I^A\,i^O$) or AB ($I^A\,I^B$).

B) The mother has Type A blood. Since there is a child with Type O blood ($i^O i^O$), the mother must have an i^O and is $I^A i^O$. Father must also have an i^O. because of the child with Type O blood. Since the other two children have Type AB and B blood respectively, father must have a B allele also. So he is $I^B i^O$.

R-8. V-3 and V-6 must have received a recessive allele from each parent. Therefore, their parents, IV-6 and IV-7, must be heterozygotes. (Note that IV-6 and IV-7 are first cousins.) One of the parents of IV-6 and IV-7 must have been heterozygote in order to pass the allele to IV-6 and IV-7. III-3 and III-4 are the parents of IV-6. III-3 came from outside the family, so it is much more probable that III-4 is heterozygous. Likewise, III-6 is more probably heterozygous than III-7, another person from outside the family. III-4 and III-6 are brother and sister. Their father II-3 is a son of I-1 who is heterozygous, so II-3 is probably the heterozygote that passed the allele to his children. Thus, the "must haves" are I-1, II-3, III-4, III-6, IV-6, IV-7.

R-9. A) Huntington's disease must be due to a dominant allele. If one parent is *Hh* (has Huntington's disease) and the other is *hh*, then half their offspring would get the *H* allele and would also have the disease. Since *H* is relatively rare, it is unlikely that the allele would be homozygous.

B) There is virtually no possibility of having a dominant condition if neither parent has the condition. Very rarely a new mutation arises that would only happen once in 10^6 to once in 10^8 times.

C) If the brother has Huntington's disease, at least one parent must have passed the *H* allele on to that brother. So at least one parent would have developed Huntington's chorea if he or she had lived long enough.

R-10. The heterozygous loci will each act like this:

Aa	X	*Aa*

Gametes: *A* and *a*   *A* and *a*

Progeny: ¼ *AA*, ½ *Aa* , and ½ *aa*

So the chance of one locus being homozygous recessive is ¼. But what about exactly one out of 20 being homozygous recessive? Here you can use the Product Rule. The chance of the one locus being heterozygous is ¼.

The chance of each of the others being other than homozygous recessive is ¾.

So the chance that one will be homozygous recessive and all the others not homozygous recessive is ¼ X ¾ X ¾ ... and so on for 19 of these ¾, or (¾)^{19}.

¼ X (¾)^{19} = 0.25 X (0.75)^{19} = 0.25 X 0.0042282826 = 0.001. Therefore, the chance that these two people will produce a child that is homozygous for exactly one of the 20 alleles the parents share is about 1 in 1000.

The chance of having an offspring that is also heterozygous for all 20 loci, like mom and dad, is (1/2)^{20}, or 0.0000009537 (about 0.000001 or 1 in 1 million).

Think of it in terms of a Punnett square. Twenty heterozygous loci mean 2^{20} different genotypes for gametes. 2^{20} is 1,048,576. The Punnett square would have 1,048,576 gamete types on each side or 1,048,576 X 1,048,576 intersections. That's about 1.1 X 10^{12} intersections. If you filled in one intersection every minute, it would take you 1.1 X 10^{12} minutes to finish. That's 1,947,830 years!

R-11. A diagram of the cross.

	EE	X	ee
P.	Dark		Light

All Ee
Intermediate

Ee X Ee

	EE	Ee	ee
F$_2$	Dark	Intermediate	Light
	1	2	1

The cross of the F1 included 6678 F2 progeny. To test the hypothesis, do a Chi-square test.

	Observed	Expected (if 1:2:1)	(0-E)	(0-E)^2	(0-E)^2/E
Dark	1605	1670.5	65.5	4290.25	2.568
Intermediate	3767	3341	420	181,476	54.318
Light	1310	1670.5	360.5	129960.25	77.792
Total	6682			Tota	134.678

Three categories is 2 degrees of freedom. The chi-square associated most closely with 134.678 is off the table. The highest chi-square on the table for 2 degrees of freedom is 9.21. So there is a extremely low probability that this deviation occurs by chance alone. As it stands we must reject our hypothesis.

R-12. Frequency of EE in this population is 3969/8070 = 0.492

Frequency of Ee in this population is 3174/8070 = 0.393
Frequency of ee in this population is 927/8070 = 0.115

So the frequency of E is all the EE and 1/2 the Ee.   0.492 + (.393/2) = 0.688
And the frequency of e is all the ee and 1/2 the Ee 0.115 + (.393/2) = 0.312

What zygote frequencies will these gamete frequencies produce.

If Hardy-Weinberg Equilibrium	Previous Generation
$p^2 = 0.688^2 = 0.473$	0.492
$2pq = 0.429 = 0.429$	0.493
$q^2 = .312^2 = 0.097$	0.115

Probably not in Hardy-Weinberg equilibrium. The ee category is much less than at equilibrium.

GLOSSARY
A PROBLEM-BASED GUIDE TO BASIC GENETICS

This glossary contains the definitions of terms students of genetics should know. These terms are printed in bold type where they are first found in the guide. Often, these terms are not formally defined in the body of the guide but are instead defined in this glossary.

Alleles   Genes governing variation of the same character that occupy corresponding positions (loci) on homologous chromosomes; alternative forms of a gene.

Anaphase   The stage of mitosis and of meiosis I and II, at which the chromosomes move to opposite poles of the cell; anaphase occurs after metaphase and before telophase.

Arm (of chromosome)   The length of a chromatid from the centromere to the end of the chromatid is its arm; thus except for chromatids with the centromere at the very end of the chromatid, the chromatid has two arms.

Autosomal gene   A gene on any chromosome other than the sex (X and Y) chromosomes.

Autosome   A chromosome other than the sex (X and Y) chromosomes.

Binomial   A mathematical expression of the kind $(p + q)^n = 1$. Used to aid in determining the probability of independently occurring events.

Binomial expansion   multiplication of a binomial by itself the number of times indicated by the exponent.

 Cell cycle   cyclic series of events in the life of a dividing eukaryotic cell; consists of mitosis, cytokinesis, and the stages of interphase; the time to complete one cell cycle is the generation time.

Centromere   Specialized constricted region of a chromatid; contains the kinetochore; in cells at prophase and metaphase, sister chromatids are joined in the vicinity of their centromeres.

Characters   The attributes of traits (See *traits*)

Chi-square   A probability distribution that is defined as $X^2 = \sum \dfrac{(O-E)^2}{E}$

Chi-square test   A statistical test used by geneticists to determine if the difference between an observed gene ratio and .the one expected because of a hypothesis is greater than would be expected on chance alone .

Chromatid   One of the two identical halves of a duplicated chromosome; the two chromatids that make up a chromosome are referred to as sister chromatids.

Chromosomes Structures in the cell nucleus that are composed of chromatin and contain the genes. The chromosomes become visible under the microscope as distinct structures during cell division.

Chromosomal sex determining mechanism   A process of male or female development directed by sex chromosomes..

Codominant alleles   Alleles in which the heterozygote expresses the characteristics of both homozygotes. Compare with *dominant allele, recessive allele* and *incompletely dominant allele.*

Coefficient   A known or constant number placed before another number by which the coefficient is multiplied. (See *exponent.*)

Complete linkage A case in which two genes only appear in parental combinations because no crossing over happens between them.

Cross A mating between two individuals.

Crossing over The process of exchange of material between homologous chromosomes.

Cross-over type A genotype in which linked genes have recombined and so are not the parental type.

Descriptive statistics Numerical analyses that summarize populations for simpler study.

Dihybrid cross A genetic cross that takes into account the behavior of alleles of two loci. Compare with *monohybrid cross.*

Diploid The condition of having two sets of chromosomes per nucleus. Compare with *haploid.*

Distribution a graphical depiction of a set of probabilities, such as a normal distribution.

Distribution, bimodal   a mode is the highest point on a normal distribution.. If there are two peaks, the distribution is bimodal.

Dominant allele An allele that is always expressed when it is present, regardless of whether it is homozygous or heterozygous. Compare with *recessive allele.*

Epistasis Condition in which certain alleles of one locus can alter the expression of genes at another locus.

Expected ratio The ratio of phenotypes or genotypes that is expected based on the hypothesis one has for the cross. See *Chi square, Chi square test, observed fre*.

Experimental design deciding what statistical methods to use while getting ready to do an experiment.

Exponent A number placed slightly above and to the right of another number The exponent indicates the power of a number

F_1 generation (first filial generation) The first generation of hybrid offspring resulting from a cross between parents from two different true-breeding lines.

F_2 generation (second filial generation) The offspring of the F_1 generation.

Fertilization Fusion of two haploid gametes; results in the formation of a 2n zygote.

First filial generation *see F_1 generation.*

Frequency The number of occurrences of an event divided by the total number of events.

G_1 phase The first gap phase within the interphase stage of the cell cycle. G_1 occurs before DNA synthesis (S-phase) begins.

G_2 phase The second gap phase within the interphase stage of the cell cycle. G_2 occurs after S-phase and before mitosis.

Gamete A sex cell; in animals and plants, an egg or sperm. In sexual reproduction the union of gametes results in the formation of a zygote.

Gamete frequency The proportion of each gamete genotype in a gene pool.

Gene pool population geneticists think in terms of a gene pool, which is a collection of all the genes in a population in their correct proportions.

Genes Hereditary factors, treated by Mendel as abstractions, but now known to be made of DNA

Genetics The science of heredity.

Genetic variation The differences between parents and offspring or among individuals of a population.

Genotype The genetic make-up of an individual. Compare with *phenotype*.

Haploid The condition of having one set of chromosomes per nucleus. Compare with *diploid*.

Hardy-Weinberg Equilibrium – the state of loci in a population when there is no mutation, selection, migration, or random mating and an infinitely large number of organism. When these five forces are not working on the populations, the gamete and zygote frequencies will not change from one generation to the next.

Hemizygous In those organisms with an XX – XY sex-determining mechanism, genes are often found on the X, but not the Y. Thus in the individuals with XY, X-linked genes are neither heterozygous nor homozygous. They are hemizygous.

Heredity The transmission of genetic information from parent to offspring.

Heterozygous Possessing a pair of unlike alleles for a particular locus. Compare with *homozygous*.

Homologous Chromosomes of the same pair in a diploid are said to be homologous.

Homozygous Possessing a pair of identical alleles for a particular locus. Compare with *heterozygous*.

Hybrid The offspring of two genetically dissimilar parents

Incomplete dominance Condition in which neither of a pair of contrasting alleles is completely expressed when the other is present.

Independent events Events in which the occurrence of one of the events has no influence on the occurrence of the other event.

Kinetochore Portions of a chromosome centromere to which the mitotic spindle fibers attach.

Linkage The tendency for a group of genes located on the same chromosome to be inherited together in successive generations.

Linked genes Genes that do not segregate independently; instead they are inherited together.

Linkage map A diagram showing the relative positions of a group of linked genes.

Locus The place on a chromosome at which the gene for a particular trait occurs, (i.e. a segment of the chromosomal DNA containing information that controls some feature of the organism); also called gene locus.

Map distance   The genetic distance between linked genes on a gene map.   Usually expressed in "map units" of 1% crossing over.

Map unit   The measure of genetic distance between gene loci on a linkage map.   One map unit is equal to 1 % crossing over.

Mathematical model – a system on paper which makes simplifying assumptions about the system to make its study easier.

Mean   a measure of central tendency in which the total measurement in the sample is divided by the sample.   This is the equation for the mean. $\bar{x} = \dfrac{\sum x}{n}$

Median   a measure of central tendency, it is the number half way from the ends of the distribution.   Half the measures are below it and half above.

Meiosis   Process in which a diploid cell undergoes two successive nuclear divisions (meiosis I and meiosis II), potentially producing four haploid nuclei; leads to the formation of gametes in animals and spores in plants.

Metaphase   The stage of mitosis, and of meiosis I and II, in which the chromosomes line up on the equatorial plane of the cell; occurs after prophase and before anaphase.

Mitosis   Division of the cell nucleus involving two daughter nuclei, each with the same number of chromosomes as the parent nucleus; mitosis consists of four phases, prophase, metaphase, anaphase, and telophase; Cytokinesis usually overlaps the telophase stage.

Migration   the movement of a population or part of a population from one region to another.

Mitotic spindle   Structure made up mostly of microtubules; provides the framework for chromosome movement during cell division.

Monohybrid cross   A genetic cross that takes into account the behavior of alleles of a single locus.   Compare to *dihybrid cross*.

Mode   a descriptive statistic, it is the most common category in a distribution/

Multiple alleles   Three or more alleles of a single locus (in a population), such as the alleles governing the ABO blood types.

Mutation   hereditary changes in genes.   A lack of mutation is one of the conditions for Hardy-Weinberg Equilibrium.

Non-parental type   In the results of a cross, those offspring that are the result of recombination and therefore do not resemble the parents.   Compare with *parental type*.

Observed ratio. The ratio that is actually seen when a cross is done

P generation (parental generation) Members of two different true-breeding lines that are crossed to produce the F_1 generation.

Parental type In the results of a cross, those offspring that did not recombine and therefore resemble the parents. Compare with *non-parental type*.

Pedigree A way of representing generations of organisms so that patterns of inheritance can be traced from one generation to the next.

Phenotype The chemical or physical appearance of an organism. Compare with *genotype*.

Population A group of organisms of the same species with defined geographical boundaries

Polygene hypothesis The idea that quantitative characters are the result of an interaction, between ba number of alleles, each with a small effect relative to the environment.

Principle of Independent Assortment Genetic principle, first noted by Gregor Mendel, that states that alleles of unlinked loci are randomly distributed to gametes.

Principle of Segregation Genetic principle, first noted by Gregor Mendel, that states that two alleles of a locus become separated into different gametes. $P = t/n$

Probability The chance that an event will occur. Found by dividing the number of occurrences of one particular event by the total number of possible events.

Product rule Rule for predicting the probability of simultaneous independent events by multiplying their individual probabilities.

Progeny The offspring of a cross.

Prophase The first stage of mitosis and of meiosis I and meiosis II. During prophase the chromosomes become visible as distinct structures, the nuclear envelope breaks down, and a spindle forms; meiotic prophase 1 is more complex and includes synapsis of homologous chromosomes and crossing over.

Punnett square Grid structure, first developed by Reginald Punnett that allows direct calculation of the probabilities of occurrence of all possible offspring of a genetic cross.

Pure-breeding Parents and offspring are the same in phenotype generation after generation.

Random mating   mating in which the choice of partners is not influenced by the genotype of the partners.   One of the   forces which are not functioning in Hardy-Weinberg equilibrium.

Recombinant type See *Non-parental type.*

Ratios The relation of two quantities; 3:1 or 9:3:3:1 are ratios as are 3/4 to 1/4, as well as 9/16 to 3/16 to 3/16 to 1/16.

Recessive allele An allele that is not expressed in the heterozygous state.   Compare to *dominant allele.*

Sample a subset of a population, the size of the sample depending on how easy it is to collect the specimens.

Sample size   the number o individuals one removes from a population for study.

Second filial generation See F_2 *generation*

Selection   the differential production of alleles in a population.   The lack of selection is one condition of Hardy-Weinberg Equilibrium..

Sex chromosome One of the chromosomes involved in sex determination.   The X and Y chromosomes in humans and *Drosophila.*

Sex determination A process by which an individual is set on the path to become male or female.

Sex-linked gene Gene carried on a sex chromosome.   In mammals, almost all the sex-linked genes are on the X chromosome (i.e., they are X-linked).

Skewed the normal curve is
Spindle See *mitotic spindle.*

Standard Deviation the square root of the variance. ,and like the variance is a measure of dispersion around the mean.

Statistical inference   Methods of making decisions about data using probability distributions. Two examples are the Chi-square test and the t-test.

Statistics   numerical techniques for measuring variable items and for drawing conclusions. Descriptive statistics describe a population, and inferential statistics allow making of conclusions about populations.

Synapsis The process of physical association of homologous chromosomes during prophase I of meiosis.

Telophase The last stage of mitosis and of meiosis I and II when, having reached the poles, chromosomes become decondensed, and a nuclear envelope forms around each group.

Test cross Genetic cross in which either an F_1 individual or an individual of unknown genotype is mated to a homozygous recessive individual.

Three-point test cross A test cross in which alleles of three gene loci are involved, often for the purpose of detecting and measuring linkage.

Trait Heritable differences between organisms.

True-breeding line A genetically pure strain of organism, (i.e., one in which all the organisms are homozygous for the gene under consideration).

Two-point test cross A test cross in which alleles of two gene loci are involved, often for the purpose of detecting and measuring linkage.

Variance `a descriptive statistic that indicates the amount of variation about the mean. It is found by using this formula: $s^2 = \dfrac{\sum(x-\bar{x})^2}{n-1}$ . .

X chromosome One of the chromosomes involved in chromosomal sex determination. It contains most of the sex-linked genes, which are called X-linked genes.

X-linked gene A gene carried on an X chromosome.

Y-chromosome One of the chromosomes involved in chromosomal sex determination. It contains only a few genes, all of which are passed from father to son.

Zygote frequency – the kinds and proportion of diploid gametes in a gene pool.

CHI-SQUARE TABLE OF PROBABILITIES

FOR CHI-SQUARE GOODNESS OF FIT TEST

df	p value											
	0.25	0.20	0.15	0.10	0.05	0.025	0.02	0.01	0.005	0.0025	0.001	0.0005
1	1.32	1.64	2.07	2.71	3.84	5.02	5.41	6.63	7.88	9.14	10.83	12.12
2	2.77	3.22	3.79	4.61	5.99	7.38	7.82	9.21	10.60	11.98	13.82	15.20
3	4.11	4.64	5.32	6.25	7.81	9.35	9.84	11.34	12.84	14.32	16.27	17.73
4	5.39	5.59	6.74	7.78	9.49	11.14	11.67	13.23	14.86	16.42	18.47	20.00
5	6.63	7.29	8.12	9.24	11.07	12.83	13.33	15.09	16.75	18.39	20.51	22.11
6	7.84	8.56	9.45	10.64	12.53	14.45	15.03	16.81	13.55	20.25	22.46	24.10
7	9.04	5.80	10.75	12.02	14.07	16.01	16.62	18.48	20.28	22.04	24.32	26.02
8	10.22	11.03	12.03	13.36	15.51	17.53	18.17	20.09	21.95	23.77	26.12	27.87
9	11.39	12.24	13.29	14.68	16.92	19.02	19.63	21.67	23.59	25.46	27.83	29.67
10	12.55	13.44	14.53	15.99	18.31	20.48	21.16	23.21	25.19	27.11	29.59	31.42
11	13.70	14.63	15.77	17.29	19.68	21.92	22.62	24.72	26.76	28.73	31.26	33.14
12	14.85	15.81	16.99	18.55	21.03	23.34	24.05	26.22	28.30	30.32	32.91	34.82
13	15.93	15.58	18.90	19.81	22.36	24.74	25.47	27.69	29.82	31.88	34.53	36.48
14	17.12	18.15	19.4	21.06	23.68	26.12	26.87	29.14	31.32	33.43	36.12	38.11
15	18.25	19.31	20.60	22.31	25.00	27.49	28.26	30.58	32.80	34.95	37.70	39.72
16	19.37	20.47	21.79	23.54	26.30	28.85	29.63	32.00	34.27	36.46	39.25	41.31
17	20.49	21.61	22.98	24.77	27.59	30.19	31.00	33.41	35.72	37.95	40.79	42.88
18	21.60	22.76	24.16	25.99	28.87	31.53	32.35	34.81	37.16	39.42	42.31	44.43
19	22.72	23.90	25.33	27.20	30.14	32.85	33.69	36.19	38.58	40.88	43.82	45.97
20	23.83	25.04	26.50	28.41	31.41	34.17	35.02	37.57	40.00	42.34	45.31	47.50
21	24.93	26.17	27.66	29.62	39.67	35.48	36.34	38.93	41.40	43.78	46.80	49.01
22	26.04	27.30	28.82	30.81	33.92	36.78	37.66	40.29	42.80	45.20	48.27	50.51
23	27.14	28.43	29.98	32.01	35.17	38.08	38.97	41.64	44.18	46.62	49.73	52.00
24	28.24	29.55	31.13	33.20	36.42	39.36	40.27	42.98	45.56	48.03	51.18	53.48
25	29.34	30.68	32.28	34.38	37.65	40.65	41.57	44.31	46.93	49.44	52.62	54.95
26	30.43	31.79	33.43	35.56	38.89	41.92	42.86	45.64	48.29	50.83	54.05	56.41
27	31.53	32.91	34.57	36.74	40.11	43.19	44.14	46.96	49.64	52.22	55.48	57.86
28	32.62	34.03	35.71	37.92	41.34	44.46	45.42	48.28	50.99	53.59	56.89	59.30
29	33.71	35.14	36.85	39.09	42.56	45.72	46.69	49.59	52.34	54.97	58.30	60.73
30	34.80	36.25	37.99	40.26	43.77	46.98	47.96	50.89	53.67	56.33	59.70	62.16
40	45.62	47.27	49.24	51.81	55.76	59.34	60.44	63.69	66.77	69.70	73.40	76.09
50	56.33	53.16	60.35	63.17	67.50	71.42	72.61	76.15	79.49	82.66	86.66	89.56
60	66.98	68.97	71.34	74.40	79.08	83.30	84.58	88.38	91.95	95.34	99.61	102.7
80	88.13	90.41	93.11	96.58	101.9	106.6	108.1	112.3	116.3	120.1	124.8	128.3
100	109.1	111.7	114.7	118.5	124.3	129.6	131.1	135.8	140.2	144.3	149.4	153.2

See instructions for use of this chi-square table in Chapter 3